AF835392

THE STANDARD ANTIQUE CLOCK VALUE GUIDE

Alex Wescot

COLLECTOR BOOKS
A Division of Schroeder Publishing Co., Inc.

Cover design: Beth Summers

COLLECTOR BOOKS
P.O. Box 3009
Paducah, Kentucky 42002-3009
www.collectorbooks.com

Contents

Introduction

Clock collecting has been a popular hobby in this country for many years. Though there have been a number of books written on the subject of clocks, this is the first comprehensive volume that identifies and places a realistic value on the most popular clocks produced by American clock companies from the 1870s to the 1930s. Prices have been averaged from actual selling prices taken from both retail galleries and auction houses.

The clocks in this guide are not museum pieces of the seventeenth or eighteenth centuries but are instead relatively common clocks produced in America in the late nineteeth-early twentieth century time period. The reader can still find these same styles and patterns at shops, antique stores, and flea markets across the country.

Eleven of the most popular and well-known makers of clocks are included here for the collector. Seth Thomas, one of the most sought-after brand of clock, is presented in detail as are the clocks of Waterbury, New Haven, Ansonia, Sessions, Ingraham, Howard, Ithaca, Gilbert, Welch, and the self-winding clocks of the Sempire Company. Along with each illustration is the type of clock, the pattern name, the approximate date, the diameter of the dial or height of the clock, and a current market value. The type or style of the clock has been referred to in the way the individual companies referred to the clocks. That is, a hanging clock may be called a wall clock, a weight clock, a hanging clock, or a regulator clock, depending on how it was referred to by the manufacturer in their ads and catalogs.

Since the cabinet styles of these companies are sometimes very similar, the collector should establish the company that built the movement before attempting to place a value on a particular clock. The prices found here are to be used as a guide. Naturally, in some parts of the country, certain clocks will be higher or lower than the values indicated. The values are for clocks in excellent running condition with refinished, restored cabinets. Clocks that do not run or have some physical flaw are worth considerably less than the value quoted.

When buying clocks that are damaged or not running, always allow cost for having the movement repaired and/or the cabinet refinished. Though the gear system of clocks is tempting work to the amateur handyman, more harm than good can result from tinkering with delicate movements by an untrained repairman. There are many experienced clock repairmen around the country who are well worth the cost of their work.

Always beware of buying replicas. The popular "nostalgia" movement has prompted many reproduction clocks that can be passed off as "old" to an unknowing collector. Sometimes such bargains cost the collector more in the long run.

SETH THOMAS CLOCKS

EXPOSITION UNIVERSELLE, HIGHEST AWARD OF MERIT

Type	Pattern	Circa	Dial	Price
Lever	Echo	1885	4½"	$75.00

Type	Pattern	Circa	Dial	Price
Lever	Echo-Alarm	1885	4½"	$95.00

Type	Pattern	Circa	Dial	Price
Lever	Echo-Calendar	1885	4½"	$145.00

Type	Pattern	Circa	Dial	Price
Lever	Elk-Alarm	1885	4½"	$120.00

Seth Thomas

Type	Pattern	Circa	Dial	Price
Lever	Joker	1885	3"	$125.00

Type	Pattern	Circa	Dial	Price
Lever	Crystal	1885	3"	$125.00

Type	Pattern	Circa	Dial	Price
Lever	Mikado	1885	3"	$135.00

Type	Pattern	Circa	Dial	Price
Lever	Artist-alarm	1885	3"	$200.00

Type	Pattern	Circa	Dial	Price
Lever	Artist	1885	3"	$195.00

Type	Pattern	Circa	Dial	Price
Lever	Lodge	1885	3"	$325.00

Type	Pattern	Circa	Dial	Price
Novelty	Dorrit	1910	2"	$120.00

Type	Pattern	Circa	Height	Price
Dresser	Night Clock "A"	1910	7¼"	$95.00

Type	Pattern	Circa	Dial	Price
Dresser	Pony	1910	2"	$125.00

Type	Pattern	Circa	Height	Price
Dresser	Traveler	1910	3½"	$95.00

Type	Pattern	Circa	Height	Price
Novelty	Cis	1910	8"	$95.00

Type	Pattern	Circa	Height	Price
Novelty	La Norma	1910	12"	$125.00

Type	Pattern	Circa	Height	Price
Novelty	Quaint	1910	9"	$175.00

Type	Pattern	Circa	Height	Price
Novelty	Ivan	1910	8¼"	$175.00

Type	Pattern	Circa	Height	Price
Novelty	Veva	1910	8½"	$195.00

Type	Pattern	Circa	Height	Price
Novelty	Dimple	1910	8¾"	$195.00

Type	Pattern	Circa	Height	Price
Novelty	LaFleur	1910	14"	$95.00

Type	Pattern	Circa	Height	Price
Dresser	Prim	1910	6"	$95.00

Type	Pattern	Circa	Dial	Price
Novelty	Holly	1910	2"	$95.00

Type	Pattern	Circa	Height	Price
Dresser	Sylvia	1910	3"	$95.00

Type	Pattern	Circa	Dial	Price
Dresser	Mode Alarm	1910	3"	$65.00

Type	Pattern	Circa	Height	Price
Dresser	Petite	1910	3½"	$120.00

Type	Pattern	Circa	Dial	Price
Novelty	School Days	1910	2"	$85.00

Type	Pattern	Circa	Dial	Price
Novelty	Tick-Tick	1910	2"	$85.00

Type	Pattern	Circa	Dial	Price
Novelty	School Girl	1910	2"	$85.00

Type	Pattern	Circa	Dial	Price
Novelty	Jacket	1910	2"	$60.00

Type	Pattern	Circa	Dial	Price
Novelty	Paddock	1910	2"	$90.00

Type	Pattern	Circa	Dial	Price
Novelty	Colin	1910	2"	$85.00

Type	Pattern	Circa	Height	Price
Novelty	Vista	1910	12¾"	$165.00

Type	Pattern	Circa	Dial	Price
Novelty	Eagle	1910	2"	$75.00

Type	Pattern	Circa	Dial	Price
Novelty	Elephant	1910	2"	$80.00

Type	Pattern	Circa	Dial	Price
Novelty	Natty	1910	2"	$75.00

Type	Pattern	Circa	Height	Price
Novelty	La Reine	1910	13½"	$125.00

Type	Pattern	Circa	Height	Price
Novelty	Duke	1910	13"	$145.00

Seth Thomas

Type	Pattern	Circa	Height	Price
Novelty	Duchess	1910	9¾"	$150.00

Type	Pattern	Circa	Height	Price
Novelty	Rex	1910	12½"	$150.00

Type	Pattern	Circa	Dial	Price
Novelty	Dainty	1910	2"	$85.00

Type	Pattern	Circa	Dial	Price
Novelty	Bouquet	1910	2"	$125.00

Type	Pattern	Circa	Height	Price
Novelty	Duchess with Lion	1910	13¾"	$185.00

Type	Pattern	Circa	Height	Price
Novelty	Lily	1910	11½"	$225.00

Type	Pattern	Circa	Height	Price
Novelty	Schiller	1910	8"	$175.00

Type	Pattern	Circa	Height	Price
Novelty	Shakespeare	1910	8"	$175.00

Type	Pattern	Circa	Height	Price
Novelty	Goethe	1910	8"	$175.00

Type	Pattern	Circa	Height	Price
Novelty	Mozart	1910	8"	$175.00

Seth Thomas

Type	Pattern	Circa	Height	Price
Novelty	Vivien	1905	27½"	$850.00

Type	Pattern	Circa	Height	Price
Novelty	Nubia	1905	18"	$425.00

Type	Pattern	Circa	Dial	Price
Pendulum	Double Dial	1905	18"	$395.00

Type	Pattern	Circa	Height	Price
Novelty	Syria	1905	16¼"	$425.00

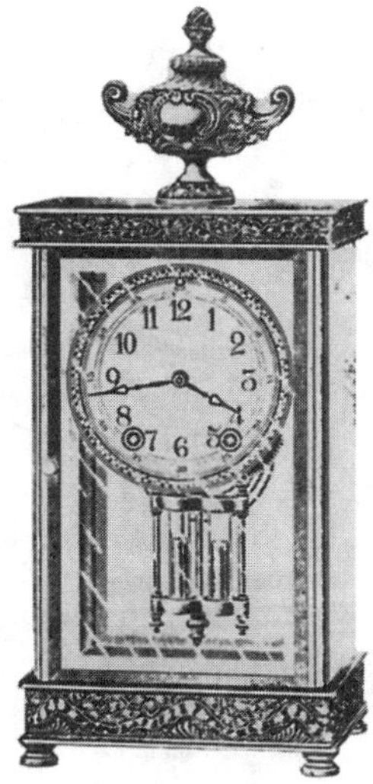

Type	Pattern	Circa	Dial	Price
Mantel	Empire No. 5	1905	3¾"	$550.00

Type	Pattern	Circa	Height	Price
Mantel	Empire No. 16	1905	16"	$750.00

Type	Pattern	Circa	Dial	Price
Mantel	Empire No. 7	1905	3¾"	$695.00

Type	Pattern	Circa	Height	Price
Mantel	Empire No. 14	1905	16"	$750.00

Type	Pattern	Circa	Height	Price
Mantel	Empire No. 12	1905	14"	$750.00

Type	Pattern	Circa	Height	Price
Mantel	Empire No. 10	1905	15¼"	$625.00

Seth Thomas

Type	Pattern	Circa	Dial	Price
Mantel	Empire No. 4	1905	3¾"	$525.00

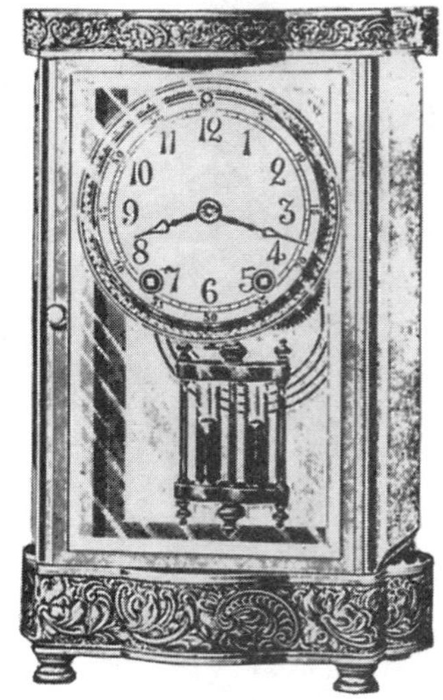

Type	Pattern	Circa	Dial	Price
Mantel	Empire No. 6	1905	3¾"	$625.00

Type	Pattern	Circa	Dial	Price
Mantel	Empire No. 40	1905	3½"	$500.00

Type	Pattern	Circa	Height	Price
Lever	Carriage No. 1	1905	5½"	$550.00

Type	Pattern	Circa	Dial	Price
Mantel	Empire No. 23	1905	4"	$525.00

Type	Pattern	Circa	Dial	Price
Mantel	Empire No. 22	1905	4"	$625.00

Type	Pattern	Circa	Dial	Price
Mantel	Orchid No. 5	1905	3¾"	$625.00

Type	Pattern	Circa	Dial	Price
Mantel	Orchid No. 6	1905	3¾"	$495.00

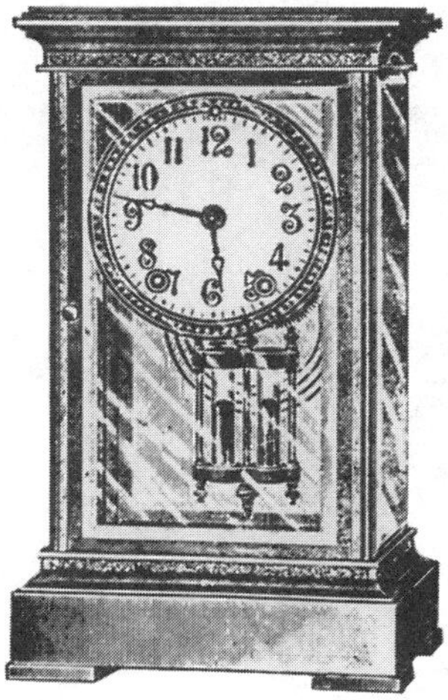

Type	Pattern	Circa	Dial	Price
Mantel	Empire No. 2	1905	3¾"	$625.00

Type	Pattern	Circa	Dial	Price
Mantel	Orchid No. 4	1905	3¾"	$325.00

Type	Pattern	Circa	Dial	Price
Mantel	Empire No. 1	1905	3¾"	$500.00

Type	Pattern	Circa	Dial	Price
Mantel	Orchid No. 3	1905	3¾"	$295.00

Seth Thomas

Type	Pattern	Circa	Dial	Price
Mantel	Zuni	1910	6"	$95.00

Type	Pattern	Circa	Dial	Price
Mantel	Aztec	1910	6"	$95.00

Type	Pattern	Circa	Dial	Price
Mantel	Onava	1910	6"	$95.00

Type	Pattern	Circa	Height	Price
Novelty	Nan	1910	7¼"	$195.00

Type	Pattern	Circa	Height	Price
Novelty	Florizel	1910	8¾"	$225.00

Type	Pattern	Circa	Height	Price
Novelty	Meditation	1910	11½"	$295.00

Type	Pattern	Circa	Dial	Price
Mantel	Cambridge	1910	6"	$275.00

Type	Pattern	Circa	Dial	Price
Mantel	New York	1910	6"	$275.00

Type	Pattern	Circa	Dial	Price
Mantel	Oxford	1910	6"	$275.00

Type	Pattern	Circa	Dial	Price
Novelty	Fountain	1910	2"	$100.00

Seth Thomas

Type	Pattern	Circa	Dial	Price
Mantel	Metals No. 6	1910	6"	$425.00

Type	Pattern	Circa	Dial	Price
Mantel	Yale	1910	6"	$275.00

Type	Pattern	Circa	Dial	Price
Mantel	Harvard	1910	6"	$275.00

Type	Pattern	Circa	Dial	Price
Mantel	Cornell	1910	6"	$275.00

Type	Pattern	Circa	Dial	Price
Mantel	Metals No. 2	1910	6"	$425.00

Type	Pattern	Circa	Dial	Price
Mantel	Metals No. 3	1910	6"	$425.00

Type	Pattern	Circa	Dial	Price
Mantel	Metals No. 4	1910	6"	$425.00

Type	Pattern	Circa	Dial	Price
Mantel	Metals No. 5	1910	6"	$450.00

Type	Pattern	Circa	Dial	Price
Mantel	Fleet No. 15	1910	6"	$295.00

Type	Pattern	Circa	Dial	Price
Mantel	Fleet No. 16	1910	6"	$295.00

Type	Pattern	Circa	Dial	Price
Mantel	Fleet No. 17	1910	6"	$295.00

Type	Pattern	Circa	Dial	Price
Mantel	Metals No. 1	1910	6"	$425.00

Type	Pattern	Circa	Dial	Price
Mantel	Viking	1905	5"	$225.00

Type	Pattern	Circa	Dial	Price
Mantel	Patmos	1905	5"	$275.00

Type	Pattern	Circa	Dial	Price
Mantel	Bosnia	1905	5"	$225.00

Type	Pattern	Circa	Dial	Price
Mantel	Pequod	1905	5"	$225.00

Type	Pattern	Circa	Dial	Price
Mantel	Texel	1905	5"	$220.00

Type	Pattern	Circa	Dial	Price
Mantel	Pasha	1905	4¼"	$225.00

Seth Thomas

Type	Pattern	Circa	Dial	Price
Mantel	Niphon	1905	5"	$225.00

Type	Pattern	Circa	Dial	Price
Mantel	Ravenna	1905	5"	$200.00

Type	Pattern	Circa	Dial	Price
Mantel	Sparta	1905	5"	$225.00

Type	Pattern	Circa	Dial	Price
Mantel	Petrel	1905	5"	$200.00

Type	Pattern	Circa	Dial	Price
Mantel	Milo	1905	5"	$225.00

Type	Pattern	Circa	Dial	Price
Mantel	Delos	1905	5"	$200.00

Type	Pattern	Circa	Dial	Price
Mantel	Toulon	1905	4¹⁄₄"	$165.00

Type	Pattern	Circa	Dial	Price
Mantel	Sussex	1905	5"	$160.00

Type	Pattern	Circa	Dial	Price
Mantel	Windsor	1905	4¹⁄₄"	$150.00

Type	Pattern	Circa	Height	Price
Mantel	Rivoli	1905	12³⁄₄"	$225.00

Type	Pattern	Circa	Dial	Price
Mantel	Elba	1905	5"	$165.00

Type	Pattern	Circa	Dial	Price
Mantel	Arno	1905	5"	$275.00

Seth Thomas

Type	Pattern	Circa	Height	Price
Weight	Parlor Calendar No. 1	1885	32"	$950.00

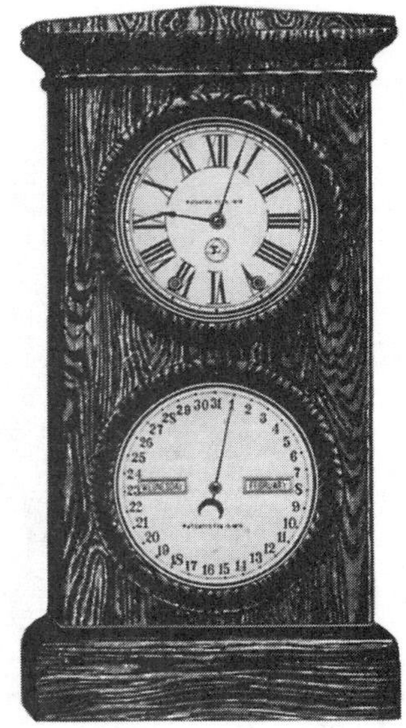

Type	Pattern	Circa	Height	Price
Spring	Parlor Calendar No. 3	1885	27"	$725.00

Type	Pattern	Circa	Height	Price
Spring	Parlor Calendar No. 4	1885	25"	$925.00

Type	Pattern	Circa	Height	Price
Spring	Parlor Calendar No. 5	1885	20"	$850.00

Type	Pattern	Circa	Height	Price
Spring	Parlor Calendar No. 6	1885	27"	$850.00

Type	Pattern	Circa	Height	Price
Spring	Parlor Calendar No. 7	1885	29"	$795.00

Type	Pattern	Circa	Height	Price
Weight	O.G.	1885	25"	$325.00

Type	Pattern	Circa	Height	Price
Weight	Column Weight	1885	25"	$325.00

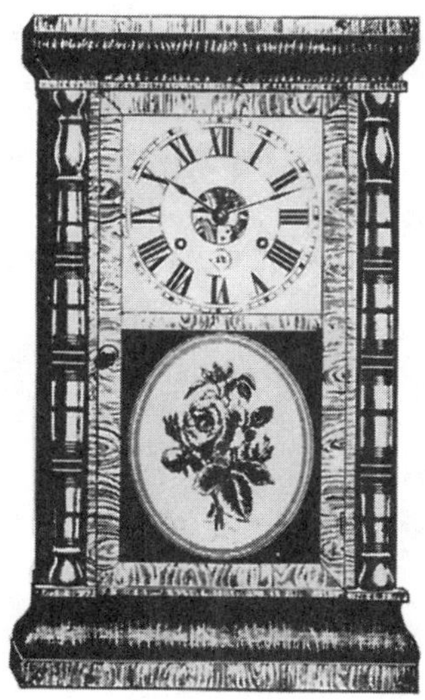

Type	Pattern	Circa	Height	Price
Weight	Column Weight	1885	25"	$425.00

Type	Pattern	Circa	Height	Price
Weight	Castle	1885	27"	$325.00

Type	Pattern	Circa	Height	Price
Weight	O.G.	1885	29½"	$395.00

Type	Pattern	Circa	Height	Price
Weight	Column	1885	32"	$450.00

Seth Thomas

Type	Pattern	Circa	Dial	Price
Spring	Carson	1885	4½"	$275.00

Type	Pattern	Circa	Height	Price
Cabinet	Normandy	1885	15"	$295.00

Type	Pattern	Circa	Dial	Price
Weight	Garfield	1885	8"	$1,250.00

Type	Pattern	Circa	Dial	Price
Weight	Lincoln	1885	8"	$925.00

Type	Pattern	Circa	Height	Price
Cabinet	Bee	1885	14"	$295.00

Type	Pattern	Circa	Height	Price
Cabinet	Sterling	1885	12"	$350.00

Type	Pattern	Circa	Dial	Price
Spring	St. Paul	1885	6"	$325.00

Type	Pattern	Circa	Dial	Price
Spring	Pittsburgh	1885	6"	$325.00

Type	Pattern	Circa	Dial	Price
Spring	Dallas	1885	4½"	$275.00

Type	Pattern	Circa	Dial	Price
Spring	Austin	1885	4½"	$300.00

Type	Pattern	Circa	Dial	Price
Spring	Atlas	1885	6"	$750.00

Type	Pattern	Circa	Dial	Price
Spring	Hecla	1885	6"	$650.00

Seth Thomas

Type	Pattern	Circa	Dial	Price
Spring	Oregon	1885	6"	$295.00

Type	Pattern	Circa	Dial	Price
Spring	Albany	1885	6"	$295.00

Type	Pattern	Circa	Dial	Price
Spring	Buffalo	1885	6"	$320.00

Type	Pattern	Circa	Dial	Price
Spring	Omaha	1885	6"	$300.00

Type	Pattern	Circa	Dial	Price
Spring	Erie	1885	6"	$350.00

Type	Pattern	Circa	Dial	Price
Spring	Greek	1885	6"	$325.00

Type	Pattern	Circa	Dial	Price
Spring	Denver	1885	6"	$295.00

Type	Pattern	Circa	Dial	Price
Spring	Athens	1885	6"	$290.00

Type	Pattern	Circa	Dial	Price
Spring	Detroit	1885	6"	$225.00

Type	Pattern	Circa	Dial	Price
Spring	Norfolk	1885	6"	$425.00

Type	Pattern	Circa	Dial	Price
Spring	Reno	1885	6"	$225.00

Type	Pattern	Circa	Dial	Price
Spring	Boston	1885	6"	$395.00

Seth Thomas

Type	Pattern	Circa	Height	Price
Spring	Albert	1885	16½"	$320.00

Type	Pattern	Circa	Height	Price
Spring	Chicago	1885	17"	$300.00

Type	Pattern	Circa	Height	Price
Spring	Nashville	1885	16"	$325.00

Type	Pattern	Circa	Height	Price
Spring	Column	1885	16"	$395.00

Type	Pattern	Circa	Height	Price
Spring	Arch Top	1885	16"	$450.00

Type	Pattern	Circa	Dial	Price
Spring	Atlanta	1885	6"	$345.00

Type	Pattern	Circa	Height	Price
Pendulum	Octagon Top	1885	9"	$250.00

Type	Pattern	Circa	Height	Price
Pendulum	Cottage	1885	9"	$225.00

Type	Pattern	Circa	Height	Price
Pendulum	Cabinet	1885	9½"	$195.00

Type	Pattern	Circa	Height	Price
Spring	Cottage	1885	14½"	$395.00

Type	Pattern	Circa	Height	Price
Spring	Round Band	1885	17"	$275.00

Type	Pattern	Circa	Height	Price
Spring	Sharp Gothic	1885	21"	$395.00

Seth Thomas

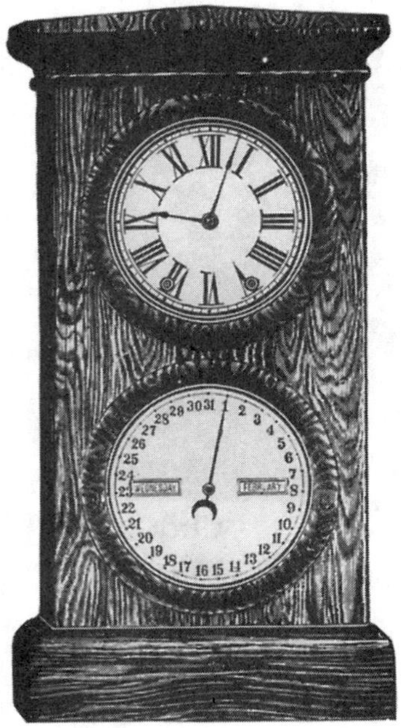

Type	Pattern	Circa	Height	Price
Spring	Parlor Calendar No. 3	1905	27"	$725.00

Type	Pattern	Circa	Height	Price
Spring	Parlor Calendar No. 5	1905	20"	$700.00

Type	Pattern	Circa	Dial	Price
Mantel	Rival	1918	5"	$295.00

Type	Pattern	Circa	Height	Price
Mantel	Wales	1905	10¾"	$300.00

Type	Pattern	Circa	Dial	Price
Mantel	Regal	1918	5"	$185.00

Type	Pattern	Circa	Dial	Price
Mantel	Mezzo	1918	5"	$150.00

Type	Pattern	Circa	Dial	Price
Mantel	Cello	1918	5"	$175.00

Type	Pattern	Circa	Dial	Price
Mantel	Alto	1918	5"	$175.00

Type	Pattern	Circa	Dial	Price
Mantel	Rex	1918	5"	$150.00

Type	Pattern	Circa	Dial	Price
Mantel	Tempo	1918	5"	$175.00

Type	Pattern	Circa	Dial	Price
Mantel	Roy	1918	5"	$150.00

Seth Thomas

Type	Pattern	Circa	Dial	Price
Lever	Engine	1885	6"	$225.00

Type	Pattern	Circa	Dial	Price
Lever	Ship's Bell	1885	6"	$395.00

Type	Pattern	Circa	Dial	Price
Lever	Banner	1885	4"	$125.00

Type	Pattern	Circa	Dial	Price
Lever	Wood Lever	1885	4"	$125.00

Type	Pattern	Circa	Dial	Price
Lever	Wood Lever	1885	10"	$265.00

Type	Pattern	Circa	Dial	Price
Lever	Chronometer	1885	3½"	$495.00

Seth Thomas

Type	Pattern	Circa	Dial	Price
Pendulum	Gallery	1905	18"	$600.00

Type	Pattern	Circa	Dial	Price
Wall	Wardroom	1905	5½"	$325.00

Type	Pattern	Circa	Dial	Price
Pendulum	Chatham	1905	12"	$300.00

Type	Pattern	Circa	Dial	Price
Lever	Navy Lever	1905	6"	$225.00

Type	Pattern	Circa	Dial	Price
Wall	No. 5009	1905	6"	$450.00

Type	Pattern	Circa	Dial	Price
Wall	No. 5020	1905	14"	$450.00

Seth Thomas

Type	Pattern	Circa	Dial	Price
Wall	Office No. 1	1918	12"	$375.00

Type	Pattern	Circa	Dial	Price
Wall	Arctic	1918	12"	$250.00

Type	Pattern	Circa	Dial	Price
Pendulum	14" Lobby	1905	14"	$160.00

Type	Pattern	Circa	Dial	Price
Wall	Chatham	1918	12"	$250.00

Type	Pattern	Circa	Dial	Price
Wall	Office No. 5	1905	12"	$140.00

Type	Pattern	Circa	Dial	Price
Lever	Engine Lever	1918	6"	$225.00

Type	Pattern	Circa	Height	Price
Spring	Drop Octagon	1885	21½"	$375.00

Type	Pattern	Circa	Height	Price
Spring	Drop Octagon	1885	23½"	$375.00

Type	Pattern	Circa	Height	Price
Spring	Drop Octagon Calendar	1885	23½"	$450.00

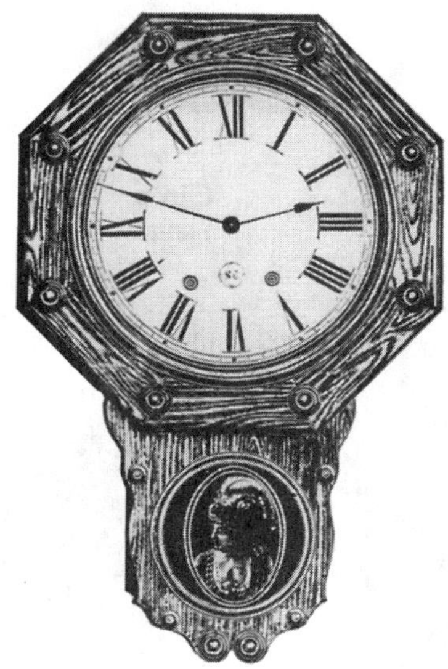

Type	Pattern	Circa	Height	Price
Spring	Office No. 2	1885	26"	$375.00

Type	Pattern	Circa	Height	Price
Spring	Office No. 1	1885	25"	$450.00

Type	Pattern	Circa	Height	Price
Spring	Signet	1885	23"	$400.00

Type	Pattern	Circa	Dial	Price
Spring	Rio	1905	12"	$495.00

Type	Pattern	Circa	Dial	Price
Spring	Brighton	1905	12"	$495.00

Type	Pattern	Circa	Dial	Price
Spring	Globe	1905	12"	$750.00

Type	Pattern	Circa	Dial	Price
Wall	Globe	1918	12"	$750.00

Type	Pattern	Circa	Dial	Price
Pendulum	18" Lobby	1905	18"	$400.00

Type	Pattern	Circa	Dial	Price
Wall	Lunar	1905	12"	$1,200.00

Type	Pattern	Circa	Dial	Price
Wall	Hudson	1918	14"	$150.00

Type	Pattern	Circa	Dial	Price
Wall	Regulator No. 2	1918	12"	$1,200.00

Type	Pattern	Circa	Dial	Price
Wall	Drop Octagon	1918	12"	$295.00+

Seth Thomas

Type	Pattern	Circa	Height	Price
Weight	Regulator No. 8	1905	56"	$8,500.00

Type	Pattern	Circa	Height	Price
Weight	Regulator No. 9	1905	48"	$3,500.00

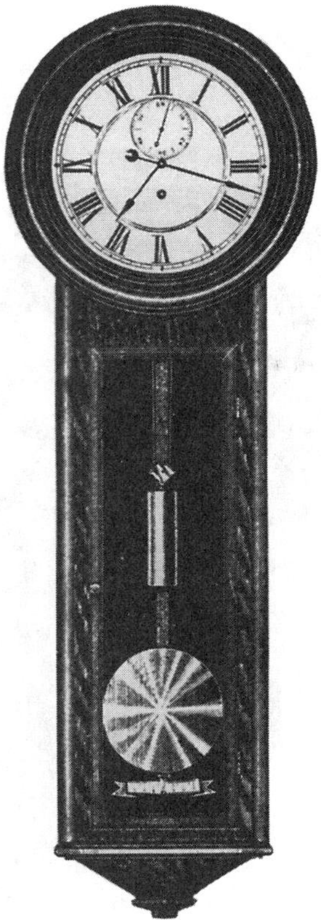

Type	Pattern	Circa	Height	Price
Weight	Regulator No. 62	1905	60"	$2,200.00+

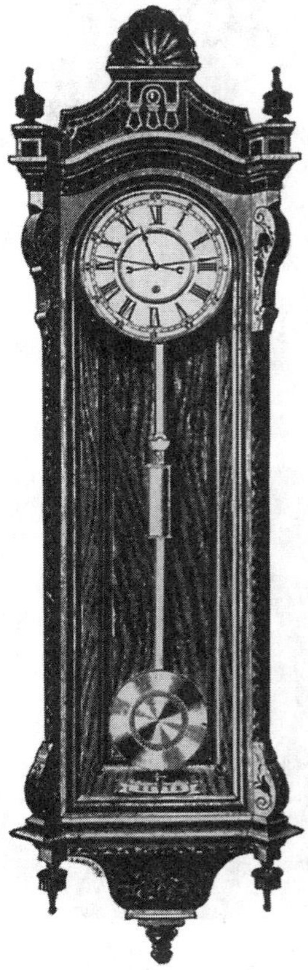

Type	Pattern	Circa	Height	Price
Weight	Regulator No. 16	1905	75"	$2,000.00+

Type	Pattern	Circa	Height	Price
Weight	Office Calendar No. 5	1885	50"	$3,500.00

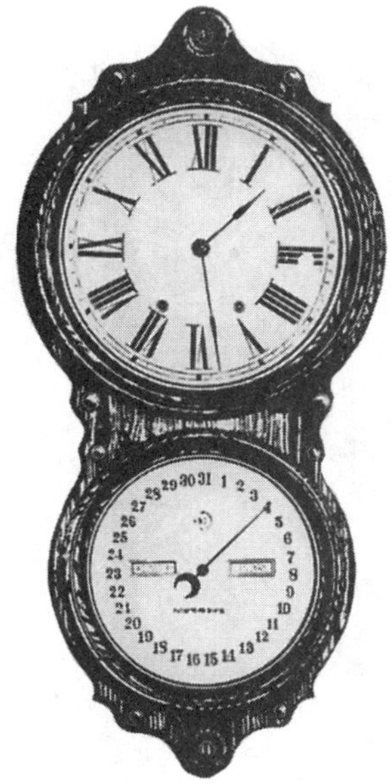

Type	Pattern	Circa	Height	Price
Spring	Office Calendar No. 6	1885	32"	$2,750.00

Type	Pattern	Circa	Height	Price
Weight	Office Calendar No. 9	1885	68"	$3,500.00

Type	Pattern	Circa	Height	Price
Weight	Office Calendar No. 8	1885	66"	$3,500.00

Seth Thomas

Type	Pattern	Circa	Height	Price
Weight	Office Calendar No. 10	1885	49"	$4,500.00

Type	Pattern	Circa	Height	Price
Weight	Office Calendar No. 11	1885	68½"	$4,000.00

Type	Pattern	Circa	Height	Price
Weight	Regulator No. 1	1885	34"	$1,200.00+

Type	Pattern	Circa	Height	Price
Weight	Regulator No. 2	1885	34"	$1,200.00+

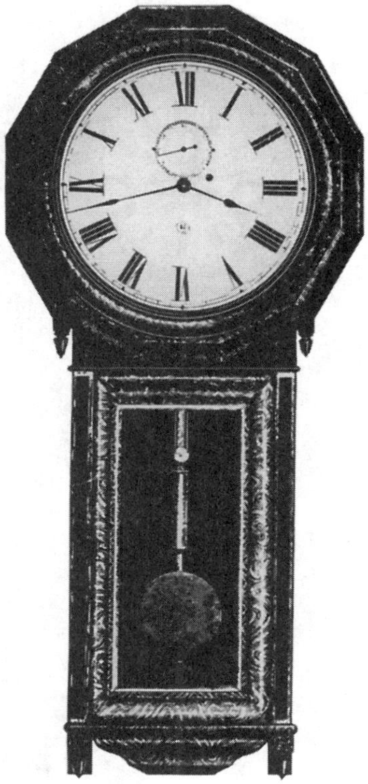

Type	Pattern	Circa	Height	Price
Weight	Regulator No. 3	1885	44"	$2,500.00

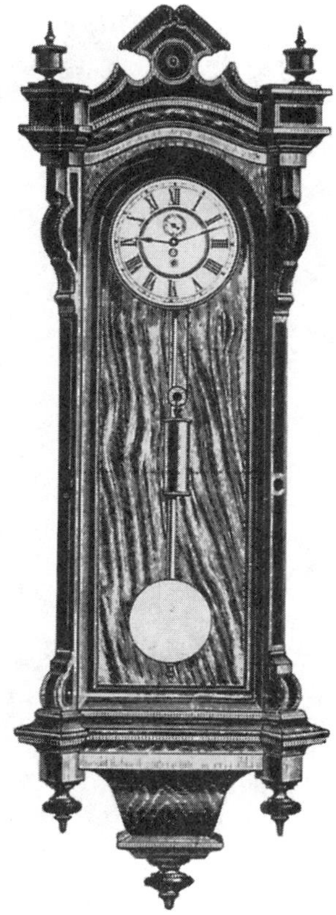

Type	Pattern	Circa	Height	Price
Weight	Regulator No. 4	1885	47"	$4,500.00

Type	Pattern	Circa	Height	Price
Weight	Regulator No. 6	1885	49"	$4,200.00

Type	Pattern	Circa	Height	Price
Weight	Regulator No. 5	1885	50"	$4,300.00

Seth Thomas

Type	Pattern	Circa	Height	Price
Weight	Regulator No. 6 Double Time	1885	49"	$6,000.00

Type	Pattern	Circa	Dial	Price
Weight	Fine Regulator No. 10	1885	14"	$3,500.00

Type	Pattern	Circa	Dial	Price
Weight	Fine Regulator No. 12	1885	14"	$3,500.00

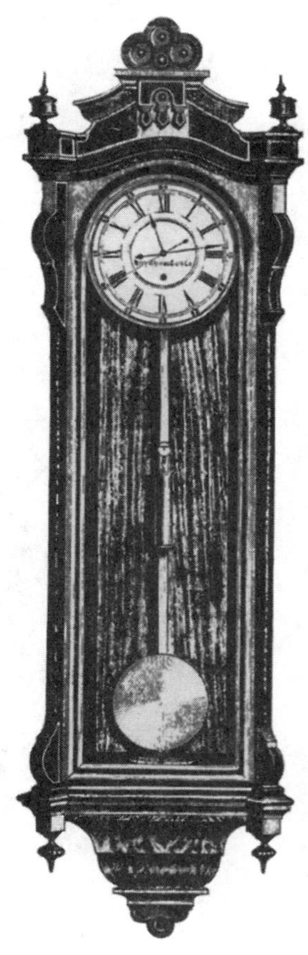

Type	Pattern	Circa	Dial	Price
Weight	Regulator No. 16	1885	12"	$3,500.00

Type	Pattern	Circa	Height	Price
Weight	Office Calendar No. 12	1905	48"	$4,500.00

Type	Pattern	Circa	Height	Price
Spring	Office Calendar No. 6	1905	32"	$4,500.00

Type	Pattern	Circa	Height	Price
Weight	Office Calendar No. 13	1905	49"	$4,000.00

Type	Pattern	Circa	Height	Price
Weight	Office Calendar No. 11	1905	68½"	$12,000.00+

Type	Pattern	Circa	Height	Price
Weight	Office Calendar No. 10	1905	49"	$4,000.00

Type	Pattern	Circa	Dial	Price
Wall	Umbria	1905	10"	$1,500.00

Type	Pattern	Circa	Height	Price
Weight	Regulator No. 6	1905	49"	$2,500.00

Type	Pattern	Circa	Height	Price
Weight	Regulator No. 6 Double Time	1905	49"	$2,200.00

Type	Pattern	Circa	Height	Price
Spring	Queen Anne	1885	36"	$1,000.00

Type	Pattern	Circa	Height	Price
Spring	Suez	1885	44"	$2,750.00

Type	Pattern	Circa	Height	Price
Weight	Flora	1885	38"	$4,200.00

Type	Pattern	Circa	Height	Price
Spring	Marcy	1885	46"	$3,500.00

Seth Thomas

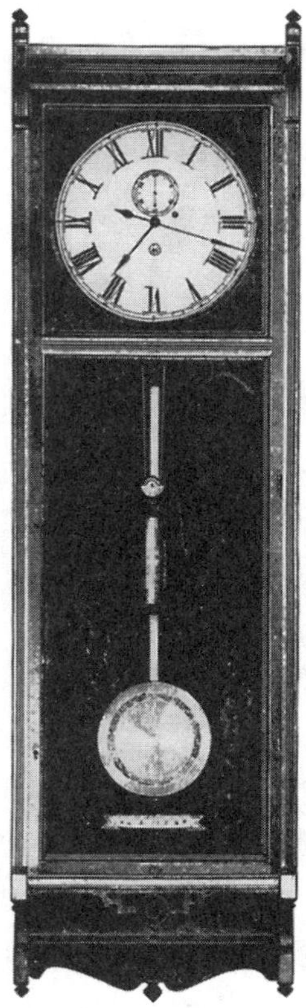

Type	Pattern	Circa	Dial	Price
Weight	Regulator No. 17	1885	14"	$2,500.00

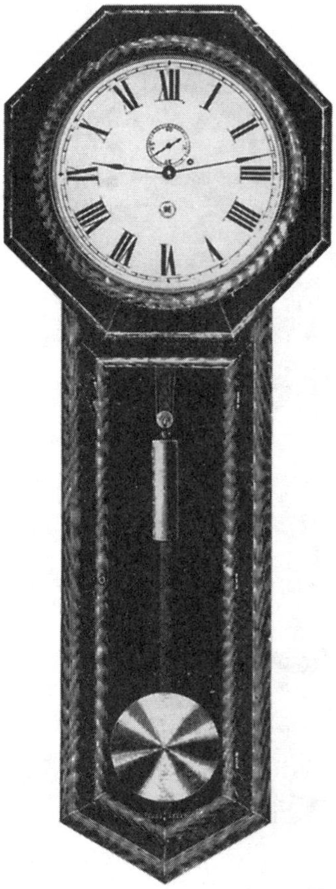

Type	Pattern	Circa	Dial	Price
Weight	Regulator No. 18	1885	14"	$3,500.00

Type	Pattern	Circa	Dial	Price
Wall	Jupiter	1905	12"	$3,200.00

Type	Pattern	Circa	Height	Price
Spring	World	1885	32"	$550.00

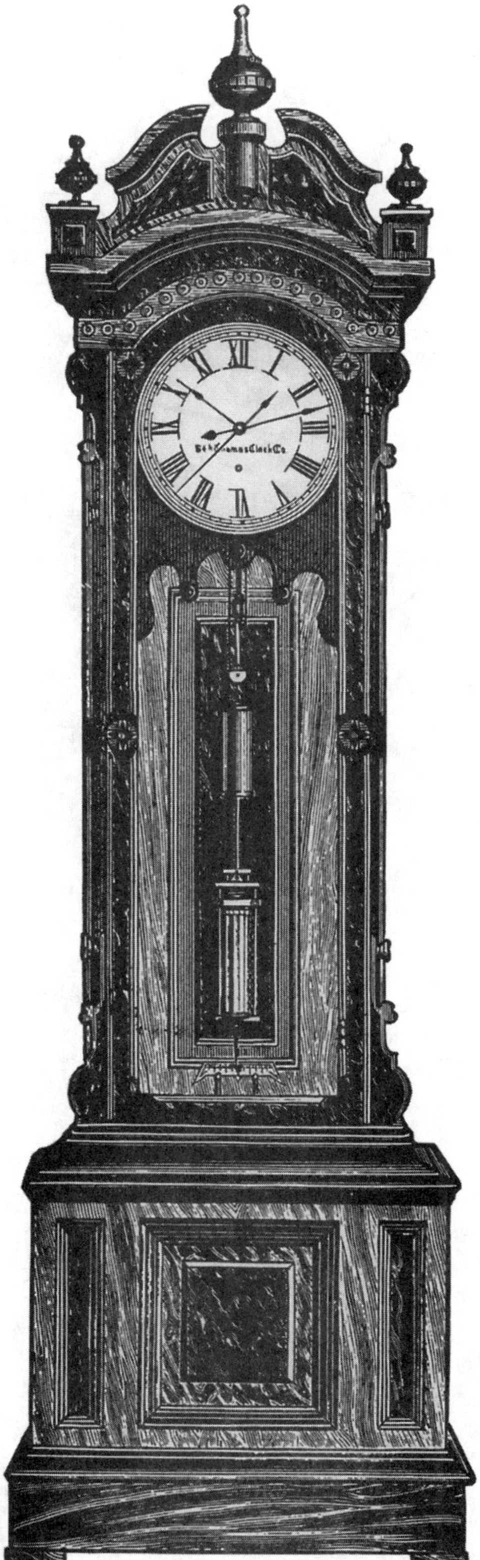

Type	Pattern	Circa	Dial	Price
Weight	Fine Regulator No. 14	1885	14"	$20,000.00

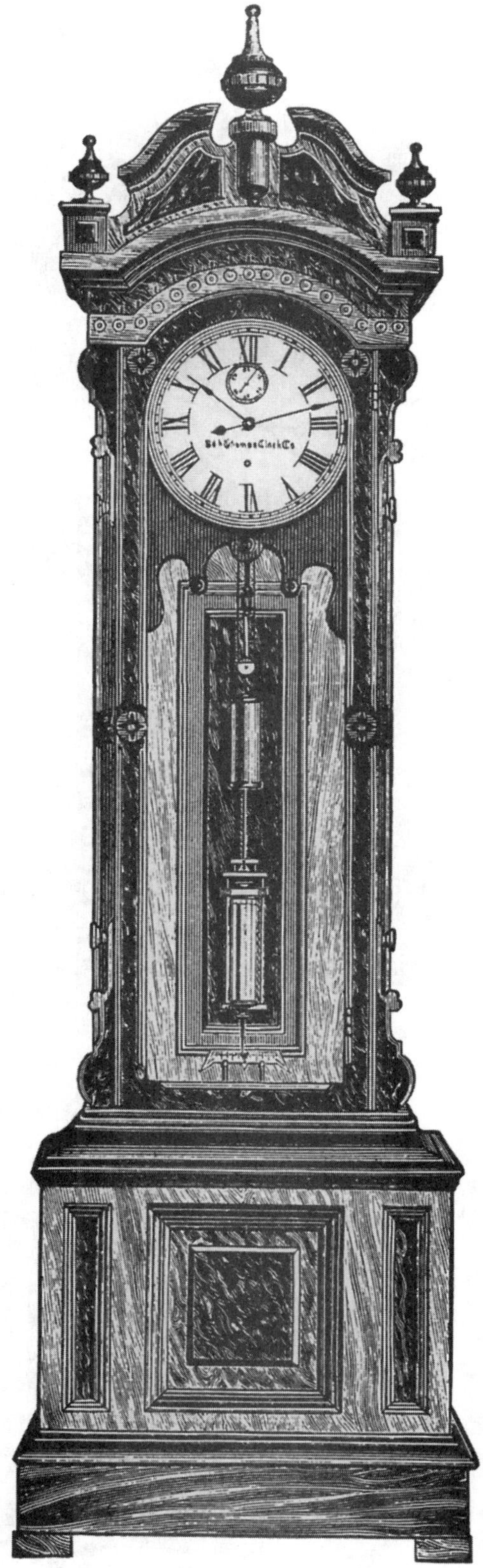

Type	Pattern	Circa	Dial	Price
Weight	Fine Regulator No. 15	1885	14"	$21,000.00

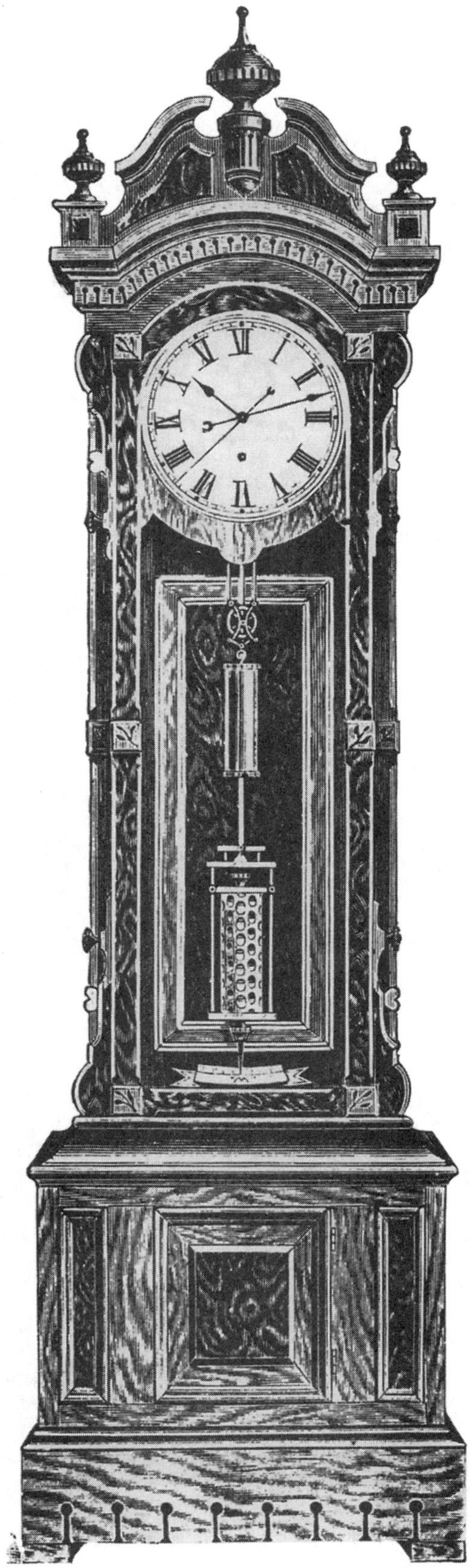

Type	Pattern	Circa	Height	Price
Weight	Fine Regulator No. 14	1905	100"	$18,000.00

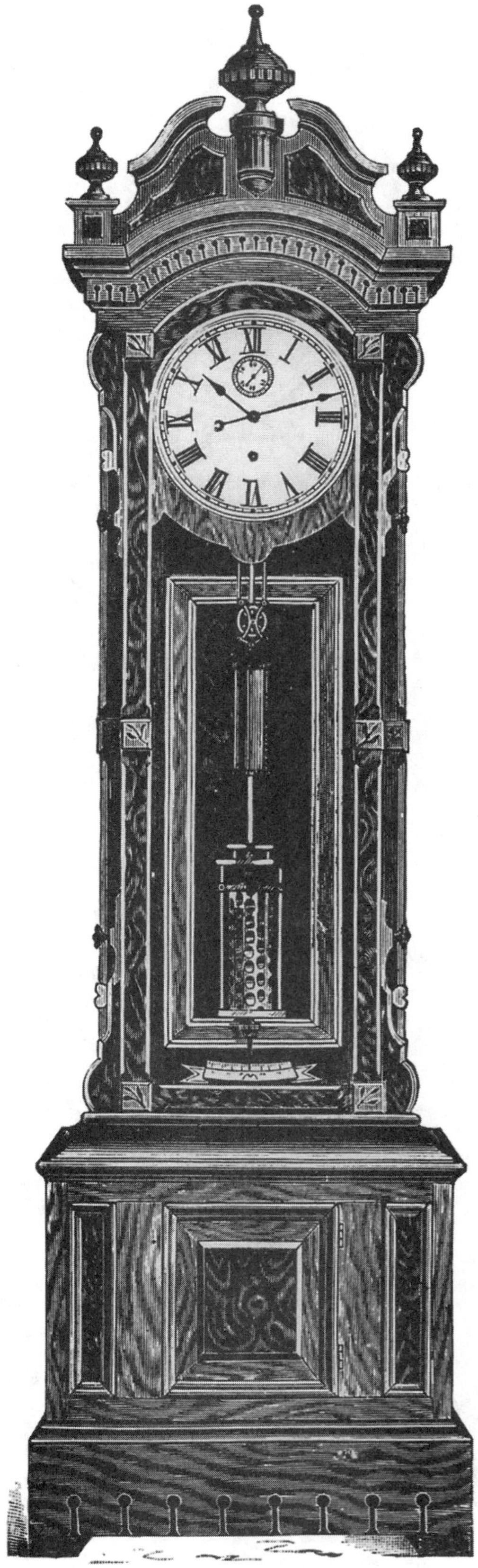

Type	Pattern	Circa	Height	Price
Weight	Fine Regulator No. 15	1905	100"	$19,000.00

Type	Pattern	Circa	Height	Price
Weight	Hall Clock No. 24	1905	93½"	$9,000.00+

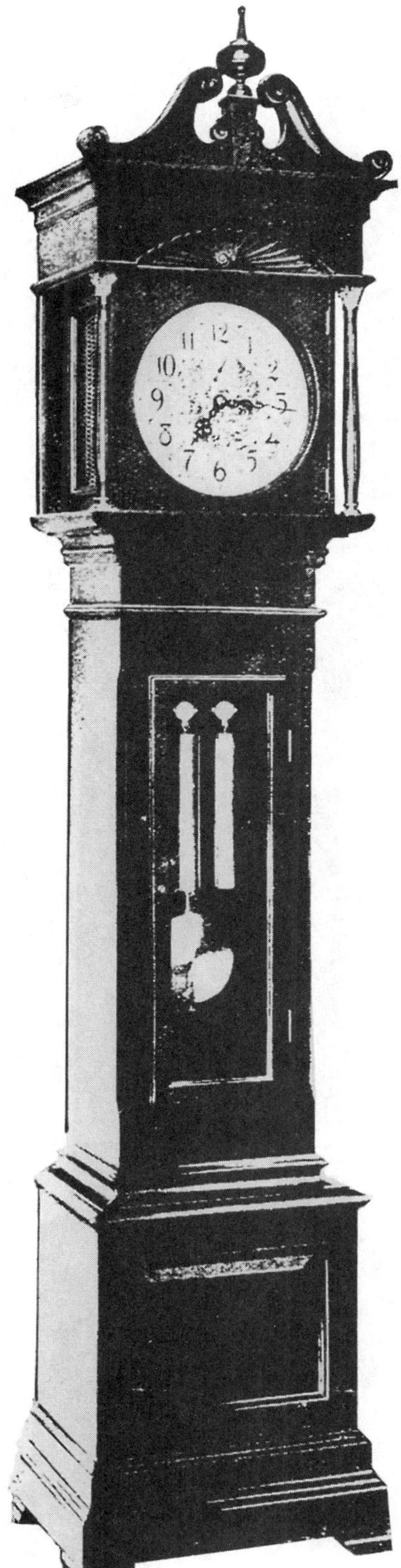

Type	Pattern	Circa	Height	Price
Weight	Hall Clock	1905	99"	$9,000.00+

Type	Pattern	Circa	Height	Price
Weight	Hall Clock No. 33	1905	97"	$12,000.00+

Type	Pattern	Circa	Height	Price
Weight	Hall Clock No. 28	1905	98"	$12,000.00+

Type	Pattern	Circa	Height	Price
Weight	Hall Clock No. 35	1905	98"	$12,000.00+

Type	Pattern	Circa	Height	Price
Weight	Hall Clock No. 34	1905	94"	$9,000.00+

Type	Pattern	Circa	Height	Price
Weight	Hall Clock No.15	1905	88"	$7,000.00+

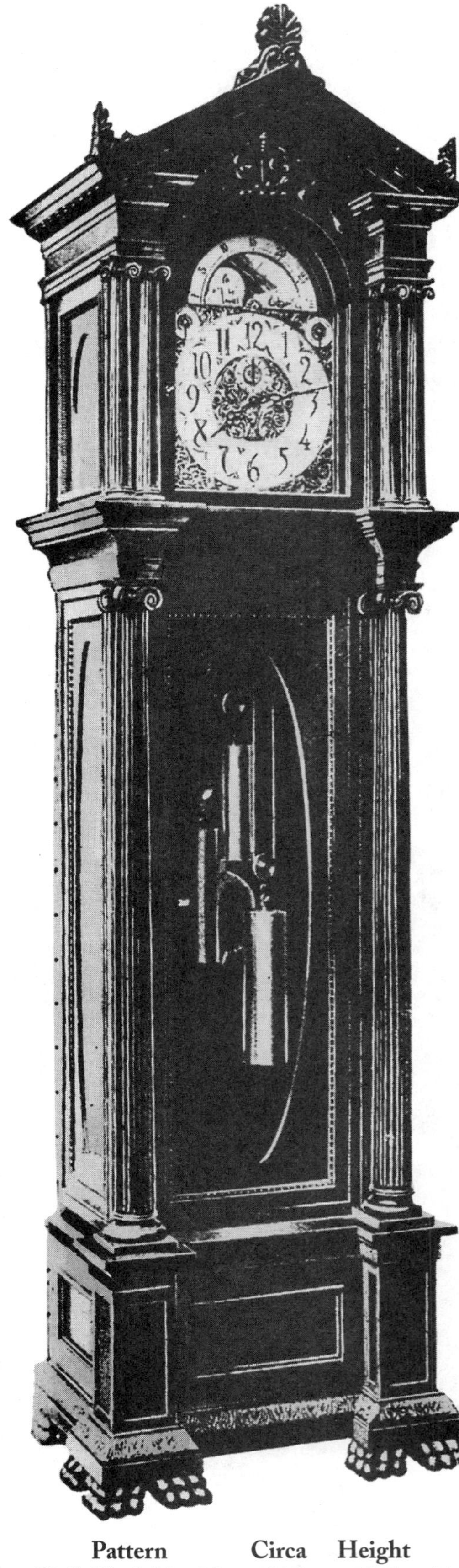

Type	Pattern	Circa	Height	Price
Weight	Hall Clock No. 38	1905	103"	$14,000.00+

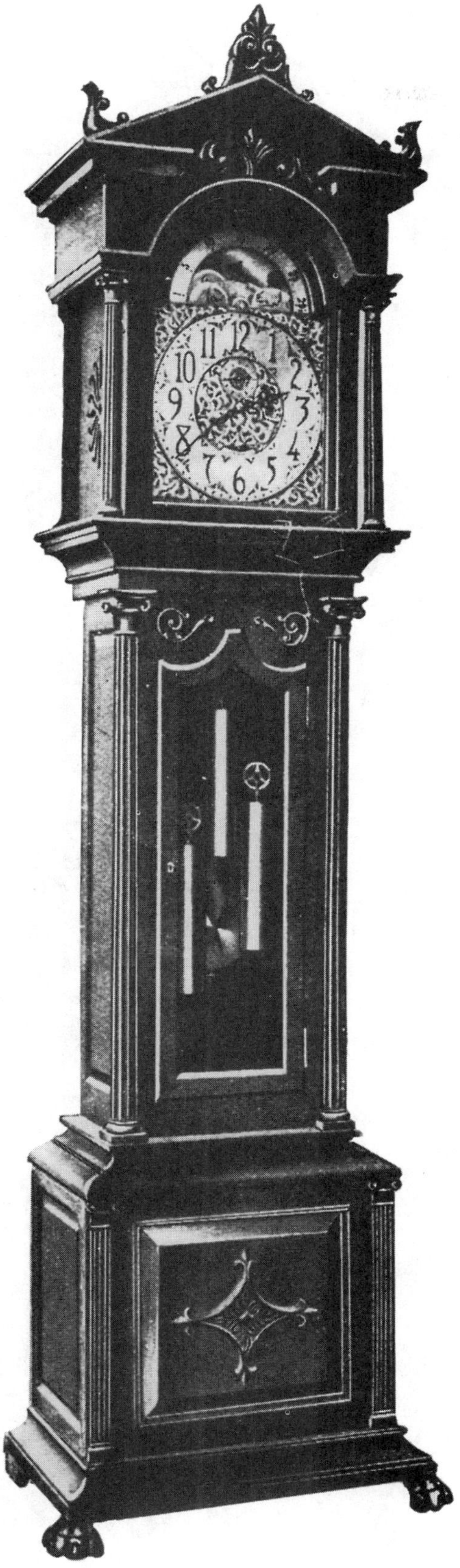

Type	Pattern	Circa	Height	Price
Weight	Hall Clock No. 37	1905	100"	$12,000.00+

Type	Pattern	Circa	Height	Price
Weight	Hall Clock No. 36	1905	103½"	$12,000.00+

Type	Pattern	Circa	Height	Price
Weight	Hall Clock No. 22	1905	98"	$10,000.00+

Type	Pattern	Circa	Height	Price
Weight	Hall Clock No. 27	1905	100"	$14,000.00+

Waterbury Clock Co.

Type	Pattern	Circa	Dial	Price
Stem-Winding	Magnet	1880	2¾"	$195.00

Type	Pattern	Circa	Dial	Price
Stem-Winding	Traveler	1880	2¾"	$195.00

Type	Pattern	Circa	Dial	Price
Stem-Winding	Sunrise-Alarm	1880	3¾"	$95.00

Type	Pattern	Circa	Dial	Price
Stem-Winding	Monitor-Calendar	1880	3¾"	$125.00

Type	Pattern	Circa	Dial	Price
Stem-Winding	Transit	1880	3¾"	$75.00

Type	Pattern	Circa	Dial	Price
Stem-Winding	Cricket Extra	1880	3"	$65.00

Type	Pattern	Circa	Dial	Price
Stem-Winding	Index-Calendar	1880	3¾"	$110.00

Type	Pattern	Circa	Dial	Price
Stem-Winding	Cricket	1880	2⅝"	$65.00

Type	Pattern	Circa	Dial	Price
Lever	Clipper	1880	4" and 6"	$120.00

Type	Pattern	Circa	Dial	Price
Lever	Clipper	1880	8"	$200.00

Type	Pattern	Circa	Dial	Price
Lever	Octagon	1880	4", 6", and 8"	$175.00

Type	Pattern	Circa	Dial	Price
Lever	Octagon	1880	10" and 12"	$250.00

Type	Pattern	Circa	Dial	Price
Lever Calendar	R.C. Octagon	1880	8" and 10"	$275.00

Waterbury

Type	Pattern	Circa	Dial	Price
Wall	Elton	1918	8"	$850.00

Type	Pattern	Circa	Dial	Price
Wall	Halifax	1918	6"	$850.00

Type	Pattern	Circa	Dial	Price
Wall	Arion	1918	12"	$325.00

Type	Pattern	Circa	Dial	Price
Wall	Bahia	1918	10"	$375.00

Type	Pattern	Circa	Dial	Price
Wall	Alpha	1930	12"	$225.00

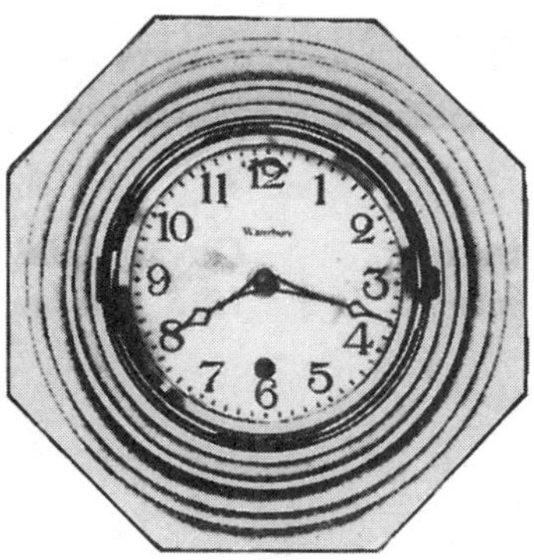

Type	Pattern	Circa	Dial	Price
Wall	Kitchen No. 4	1930	6"	$95.00

Type	Pattern	Circa	Dial	Price
Wall	Kitchen No. 3	1930	6"	$85.00

Type	Pattern	Circa	Dial	Price
Wall	Drop Octagon	1918	12"	$375.00

Type	Pattern	Circa	Dial	Price
Wall	Andes	1918	12"	$295.00

Waterbury

Type	Pattern	Circa	Dial	Price
Cabinet	Lowell	1918	6"	$75.00

Type	Pattern	Circa	Dial	Price
Cabinet	Carolina	1918	5½"	$65.00

Type	Pattern	Circa	Dial	Price
Cabinet	Stanhope	1918	6"	$65.00

Type	Pattern	Circa	Dial	Price
Cabinet	Windsor	1918	6"	$75.00

Type	Pattern	Circa	Dial	Price
Cabinet	Kent	1918	6"	$65.00

Type	Pattern	Circa	Dial	Price
Cabinet	Croyden	1918	5"	$65.00

Type	Pattern	Circa	Dial	Price
Cabinet	Exeter	1918	5"	$65.00

Type	Pattern	Circa	Dial	Price
Novelty	Alice	1918	2"	$65.00

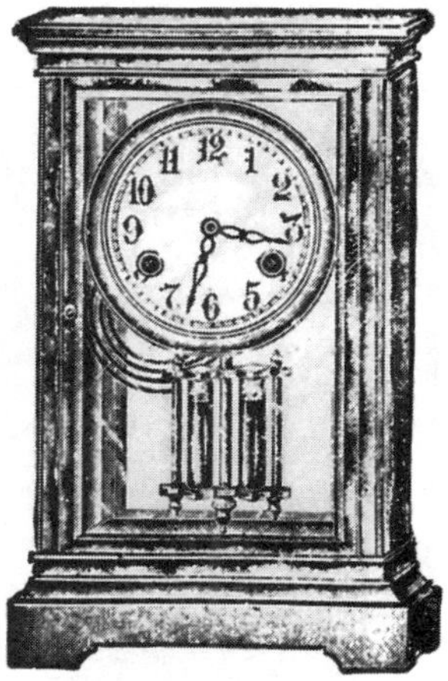

Type	Pattern	Circa	Dial	Price
Cabinet	Brest Clock	1918	3½"	$450.00

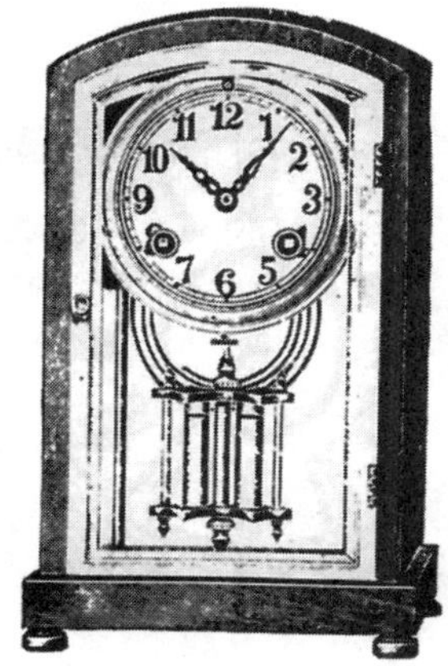

Type	Pattern	Circa	Dial	Price
Cabinet	Trouville Clock	1918	3½"	$375.00

Type	Pattern	Circa	Dial	Price
Cabinet	Loirol Clock	1918	5"	$245.00

Type	Pattern	Circa	Dial	Price
Cabinet	Vercelle Clock	1918	5"	$245.00

Type	Pattern	Circa	Dial	Price
Cabinet	Landes	1918	4¼"	$450.00

Type	Pattern	Circa	Dial	Price
Cabinet	Loiret	1918	4½"	$395.00

Waterbury

Type	Pattern	Circa	Dial	Price
Wall	Arion	1918	10"	$375.00

Type	Pattern	Circa	Dial	Price
Wall	Heron	1918	10"	$750.00

Type	Pattern	Circa	Dial	Price
Wall	Admiral	1918	12"	$795.00

Type	Pattern	Circa	Dial	Price
Wall	Crane	1918	12"	$695.00

Type	Pattern	Circa	Dial	Price
Novelty	Tambour Alarm No. 1	1918	4"	$120.00

Type	Pattern	Circa	Dial	Price
Novelty	Agout	1918	2"	$95.00

Type	Pattern	Circa	Dial	Price
Novelty	White Novelty No. 4	1918	2"	$90.00

Type	Pattern	Circa	Dial	Price
Novelty	Alene	1918	2"	$95.00

Type	Pattern	Circa	Dial	Price
Cabinet	Winston	1918	5"	$90.00

Type	Pattern	Circa	Dial	Price
Cabinet	Wellington	1918	6"	$90.00

Waterbury

Type	Pattern	Circa	Dial	Price
Wall	Admiral	1930	12"	$925.00

Type	Pattern	Circa	Dial	Price
Wall	No. 9117	1930	12"	$600.00

Type	Pattern	Circa	Dial	Price
Wall	Irving	1930	12"	$120.00

Type	Pattern	Circa	Dial	Price
Wall	Avon	1930	7"	$95.00

Type	Pattern	Circa	Dial	Price
Alarm	Relay Jr.	1930	2¼"	$65.00

Type	Pattern	Circa	Dial	Price
Alarm	Relay	1930	4"	$65.00

Type	Pattern	Circa	Dial	Price
Alarm	Trusty	1930	4"	$90.00

Type	Pattern	Circa	Dial	Price
Alarm	Sunrise	1930	4"	$90.00

Waterbury

Type	Pattern	Circa	Dial	Price
Wall	No. 9122	1930	6"	$475.00

Type	Pattern	Circa	Dial	Price
Wall	No. 9124	1930	8"	$195.00

Type	Pattern	Circa	Dial	Price
Wall	No. 9123	1930	6"	$475.00

Type	Pattern	Circa	Dial	Price
Wall	Pelican	1930	12"	$550.00

Type	Pattern	Circa	Dial	Price
Kitchen	No. 9090	1930	6"	$375.00

Type	Pattern	Circa	Dial	Price
Kitchen	No. 9091	1930	6"	$375.00

Type	Pattern	Circa	Dial	Price
Kitchen	No. 9092	1930	6"	$375.00

Type	Pattern	Circa	Dial	Price
Kitchen	No. 9093	1930	6"	$375.00

Waterbury

Type	Pattern	Circa	Dial	Price
Mantel	Adley	1918	2½"	$75.00

Type	Pattern	Circa	Dial	Price
Mantel	Agra	1918	2½"	$75.00

Type	Pattern	Circa	Dial	Price
Novelty	Tambour Alarm No. 2	1918	4"	$75.00

Type	Pattern	Circa	Dial	Price
Novelty	Acorn	1918	2"	$75.00

Type	Pattern	Circa	Dial	Price
Mantel	Adams	1918	3½"	$75.00

Type	Pattern	Circa	Dial	Price
Mantel	Aladdin	1918	3½"	$90.00

Type	Pattern	Circa	Dial	Price
Mantel	Abington	1918	3½"	$90.00

Type	Pattern	Circa	Dial	Price
Mantel	Athlone	1918	3½"	$75.00

Type	Pattern	Circa	Dial	Price
Wall	Breton	1918	8"	$2,400.00

Type	Pattern	Circa	Dial	Price
Wall	Fostoria	1918	12"	$2,600.00

Type	Pattern	Circa	Dial	Price
Wall	Regulator No. 18	1918	12"	$2,200.00

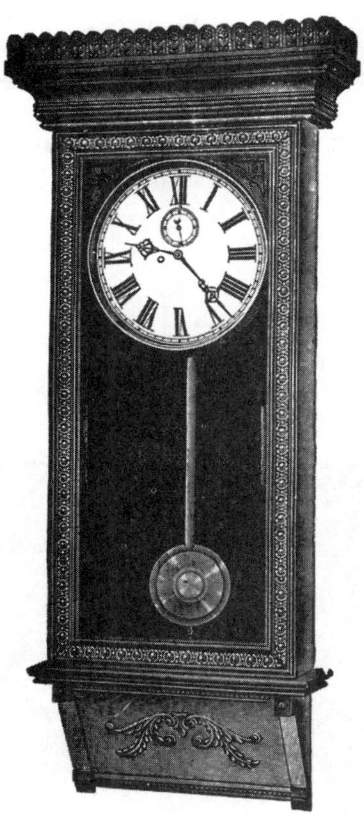

Type	Pattern	Circa	Dial	Price
Wall	Regulator No. 67	1918	12"	$2,700.00

Type	Pattern	Circa	Dial	Price
Mantel	Dover	1918	5½"	$265.00

Type	Pattern	Circa	Dial	Price
Mantel	Duluth	1918	5½"	$250.00

Type	Pattern	Circa	Dial	Price
Mantel	Delano	1918	5½"	$275.00

Type	Pattern	Circa	Dial	Price
Mantel	Dwyer	1918	5½"	$300.00

Type	Pattern	Circa	Dial	Price
Mantel	Dunstable	1918	5½"	$275.00

Type	Pattern	Circa	Dial	Price
Mantel	Dorset	1918	5½"	$225.00

Type	Pattern	Circa	Dial	Price
Mantel	Albert	1918	2½"	$75.00

Type	Pattern	Circa	Dial	Price
Mantel	Elmira	1918	5"	$75.00

Type	Pattern	Circa	Dial	Price
Mantel	Ajax	1918	2½"	$75.00

Type	Pattern	Circa	Dial	Price
Mantel	White Novelty No. 5	1918	2½"	$65.00

Type	Pattern	Circa	Dial	Price
Mantel	Alaska	1918	2½"	$65.00

Type	Pattern	Circa	Dial	Price
Novelty	Abner	1918	2"	$65.00

Waterbury

Type	Pattern	Circa	Dial	Price
Automobile	Cowl Dash No. 1	1918	2½"	$175.00

Type	Pattern	Circa	Dial	Price
Automobile	Florida	1918	2½"	$195.00

Type	Pattern	Circa	Dial	Price
Mantel	Conductor	1918	2"	$300.00

Type	Pattern	Circa	Dial	Price
Lever	Oak Lever	1918	4"	$125.00

Type	Pattern	Circa	Dial	Price
Novelty	Leather Novelty No. 2	1918	2"	$65.00

Type	Pattern	Circa	Dial	Price
Novelty	Leather Novelty No. 4	1918	2"	$65.00

Type	Pattern	Circa	Dial	Price
Mantel	Emporia	1918	6"	$85.00

Type	Pattern	Circa	Dial	Price
Mantel	Edmond	1918	5½"	$125.00

Type	Pattern	Circa	Dial	Price
Mantel	Eugenia	1918	5½"	$95.00

Type	Pattern	Circa	Dial	Price
Mantel	Eagle	1918	5½"	$125.00

Type	Pattern	Circa	Dial	Price
Mantel	Byron	1918	6"	$75.00

Type	Pattern	Circa	Dial	Price
Mantel	Englewood	1918	5"	$75.00

Waterbury

Type	Pattern	Circa	Dial	Price
Mantel	Espy	1918	5½"	$250.00

Type	Pattern	Circa	Dial	Price
Mantel	Edmond	1918	5½"	$225.00

Type	Pattern	Circa	Dial	Price
Mantel	Duarte	1918	5½"	$250.00

Type	Pattern	Circa	Dial	Price
Mantel	Delano	1918	5½"	$250.00

Type	Pattern	Circa	Dial	Price
Mantel	Delmonte	1918	5½"	$250.00

Type	Pattern	Circa	Dial	Price
Mantel	Dorset	1918	5½"	$250.00

Type	Pattern	Circa	Dial	Price
Regulator	Toronto	1880	8"	$2,400.00

Type	Pattern	Circa	Dial	Price
Regulator	Quebec	1880	8"	$2,500.00

Type	Pattern	Circa	Dial	Price
Regulator	Ontario	1880	8"	$3,250.00

Type	Pattern	Circa	Dial	Price
Regulator	Waterbury Regulator, No. 2	1880	12"	$2,600.00

Type	Pattern	Circa	Dial	Price
Regulator	Waterbury Regulator, No. 1	1880	12"	$2,200.00

Type	Pattern	Circa	Dial	Price
Regulator	Camden	1880	10"	$2,600.00

Type	Pattern	Circa	Dial	Price
Regulator	Berlin	1880	10"	$2,500.00

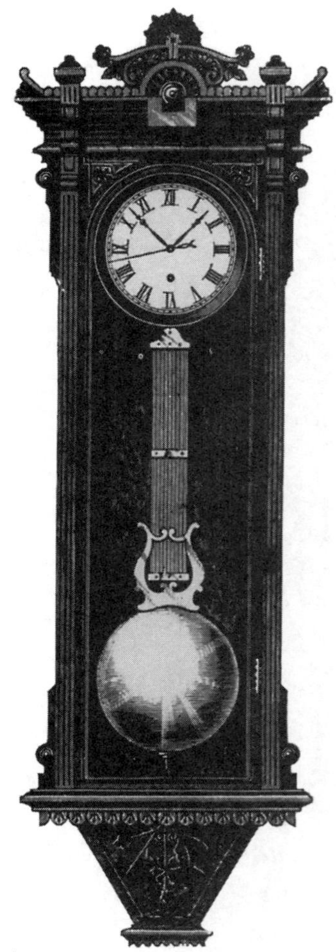

Type	Pattern	Circa	Dial	Price
Regulator	Walnut Regulator, No. 9	1880	12"	$4,000.00

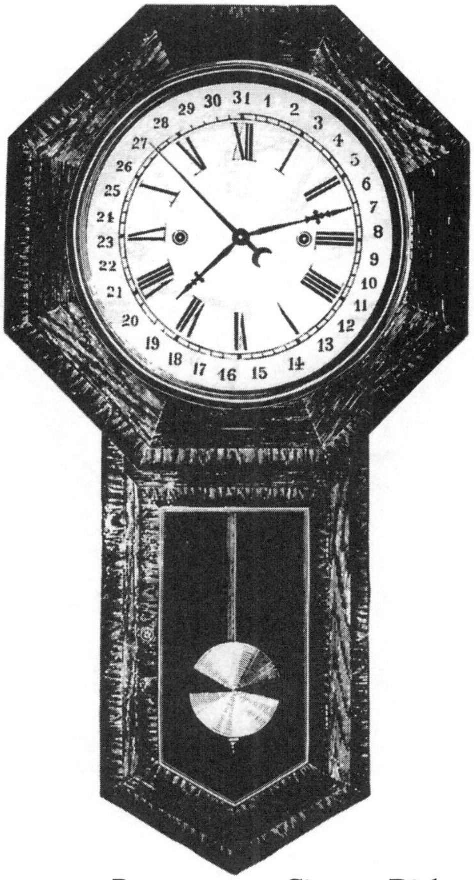

Type	Pattern	Circa	Dial	Price
Regulator	Regent-Calendar	1880	12"	$675.00

Type	Pattern	Circa	Dial	Price
Regulator	Regent	1880	12"	$500.00

Type	Pattern	Circa	Dial	Price
Regulator	Montreal	1880	8"	$750.00

Type	Pattern	Circa	Dial	Price
Regulator	Kingston	1880	8"	$750.00

Waterbury

Type	Pattern	Circa	Dial	Price
Regulator	Drop Octagon, R.C.	1880	12"	$385.00

Type	Pattern	Circa	Dial	Price
Regulator	Glenwood	1880	12"	$375.00

Type	Pattern	Circa	Dial	Price
Regulator	Drop Octagon, R.C. Calendar	1880	12"	$425.00

Type	Pattern	Circa	Dial	Price
Regulator	Glenwood, Calendar	1880	12"	$395.00

Type	Pattern	Circa	Height	Price
Weight	Courier	1880	25⅞"	$350.00

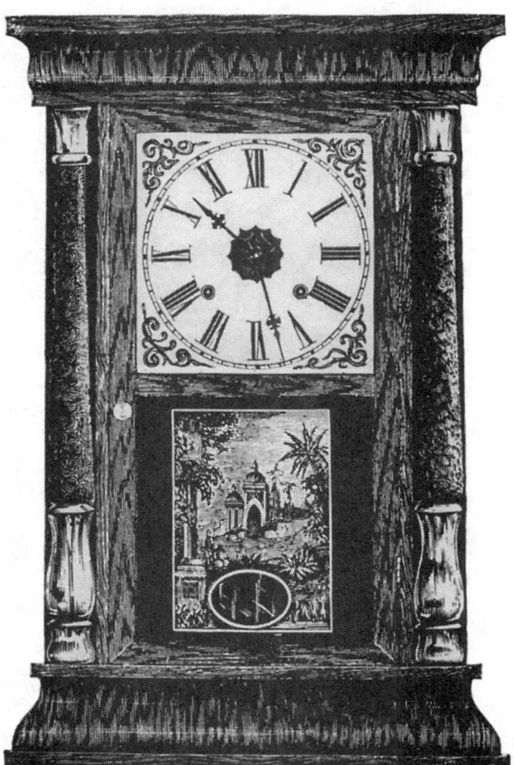

Type	Pattern	Circa	Height	Price
Weight	Column	1880	25"	$375.00

Type	Pattern	Circa	Height	Price
Weight	O.O.G.	1880	30"	$395.00

Type	Pattern	Circa	Dial	Price
Regulator	Drop Octagon, Gilt, Calendar	1880	10"	$450.00

Waterbury

Type	Pattern	Circa	Height	Price
Mantel	Sultan, No. 1, V.P.	1880	17⅛"	$295.00

Type	Pattern	Circa	Height	Price
Weight	O.G.	1880	25⅞"	$350.00

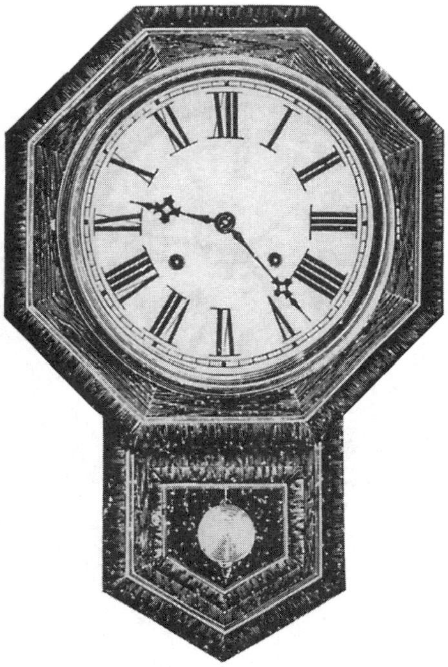

Type	Pattern	Circa	Dial	Price
Regulator	Drop Octagon, Gilt	1880	10"	$375.00

Type	Pattern	Circa	Height	Price
Weight	O.O.G.	1880	26"	$350.00

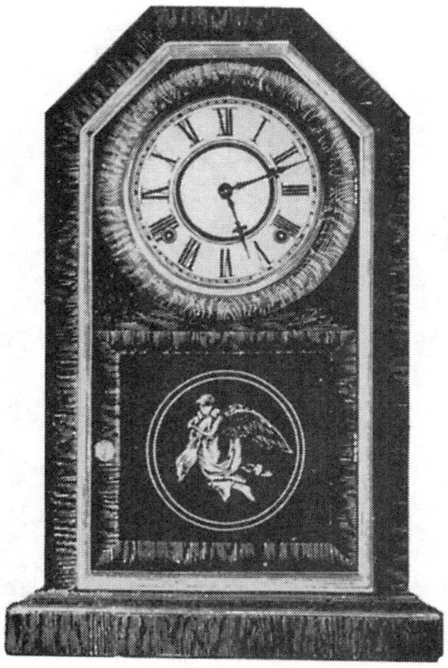

Type	Pattern	Circa	Height	Price
Mantel	Celtic	1880	17⅞"	$350.00

Type	Pattern	Circa	Height	Price
Mantel	Sultan, No. 1	1880	17⅛"	$350.00

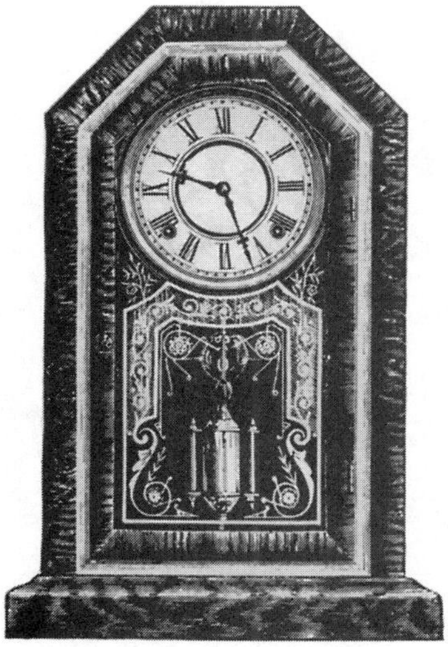

Type	Pattern	Circa	Height	Price
Mantel	Celtic, V.P.	1880	17⅞"	$350.00

Type	Pattern	Circa	Height	Price
Mantel	Sultan, No. 1	1880	17⅛"	$365.00

Type	Pattern	Circa	Height	Price
Mantel	Chester	1880	16⅞"	$275.00

Type	Pattern	Circa	Height	Price
Mantel	Sultan, No. 1, Extra	1880	17⅛"	$365.00

Waterbury

Type	Pattern	Circa	Height	Price
Mantel	Round Gothic	1880	19"	$275.00

Type	Pattern	Circa	Height	Price
Mantel	Cardinal, V.P.	1880	16¼"	$325.00

Type	Pattern	Circa	Height	Price
Mantel	Corinth	1880	16"	$350.00

Type	Pattern	Circa	Height	Price
Mantel	Vulcan	1880	17"	$350.00

Type	Pattern	Circa	Height	Price
Mantel	Cardinal	1880	16¼"	$350.00

Type	Pattern	Circa	Height	Price
Mantel	Vulcan	1880	17"	$350.00

Type	Pattern	Circa	Height	Price
Mantel	Sharp Gothic	1880	19½"	$275.00

Type	Pattern	Circa	Height	Price
Mantel	Column Spring	1880	16"	$275.00

Type	Pattern	Circa	Height	Price
Mantel	Sharp Gothic Extra	1880	19½"	$275.00

Type	Pattern	Circa	Height	Price
Mantel	Florence	1880	17¼"	$275.00

Type	Pattern	Circa	Height	Price
Mantel	Sultan, No. 2	1880	15"	$275.00

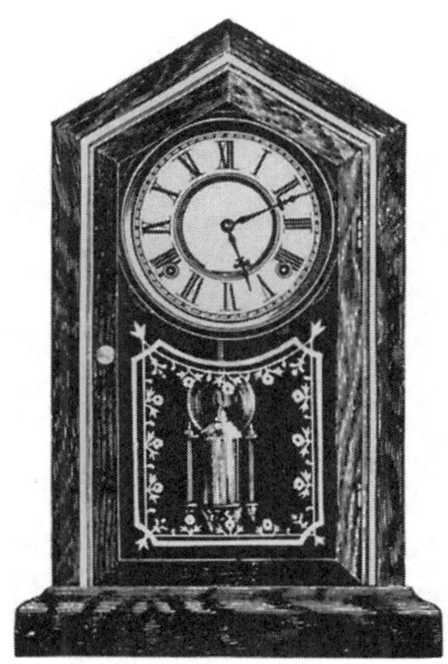

Type	Pattern	Circa	Height	Price
Mantel	Florence, V.P.	1880	17¼"	$275.00

Waterbury

Type	Pattern	Circa	Height	Price
Mantel	Beacon Time	1880	12"	$50.00

Type	Pattern	Circa	Height	Price
Mantel	Prince	1880	12"	$300.00

Type	Pattern	Circa	Height	Price
Mantel	Beacon No. 1	1880	17"	$180.00
Mantel	Beacon No. 2	1880	15"	$165.00

Type	Pattern	Circa	Height	Price
Mantel	O.O.G. Spring	1880	18"	$325.00

Type	Pattern	Circa	Height	Price
Mantel	Sultan Time	1880	12"	$300.00

Type	Pattern	Circa	Height	Price
Mantel	Courier Spring	1880	17½"	$275.00

Type	Pattern	Circa	Height	Price
Mantel	Tick-Tick	1880	10½"	$145.00

Type	Pattern	Circa	Height	Price
Mantel	Cottage Extra	1880	12½"	$150.00

Type	Pattern	Circa	Height	Price
Mantel	Cottage No. 2	1880	12"	$150.00

Type	Pattern	Circa	Height	Price
Mantel	Small Gothic	1880	14½"	$275.00

Type	Pattern	Circa	Height	Price
Mantel	Cottage Extra	1880	11½"	$150.00

Type	Pattern	Circa	Height	Price
Mantel	Small Gothic Extra	1880	14⅞"	$275.00

Waterbury

Type	Pattern	Circa	Dial	Price
Mantel	Melrose	1880	6"	$295.00

Type	Pattern	Circa	Dial	Price
Mantel	Irving	1880	6"	$325.00

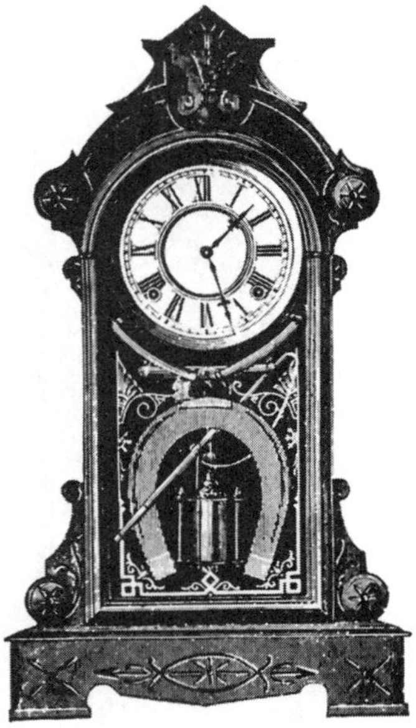

Type	Pattern	Circa	Dial	Price
Mantel	Durham	1880	6"	$450.00

Type	Pattern	Circa	Dial	Price
Mantel	Hecla	1880	6"	$595.00

Type	Pattern	Circa	Dial	Price
Mantel	Galway	1880	8"	$850.00

Type	Pattern	Circa	Dial	Price
Mantel	Ridgewood	1880	4"	$300.00

Type	Pattern	Circa	Dial	Price
Mantel	Portland	1880	5"	$300.00

Type	Pattern	Circa	Dial	Price
Mantel	Croton	1880	4"	$375.00

Type	Pattern	Circa	Dial	Price
Mantel	Belfast	1880	5"	$375.00

Type	Pattern	Circa	Dial	Price
Mantel	Lowell	1880	5"	$375.00

Type	Pattern	Circa	Dial	Price
Mantel	Amboy	1880	5"	$375.00

Waterbury

Type	Pattern	Circa	Dial	Price
Mantel	Salem	1880	5"	$350.00

Type	Pattern	Circa	Dial	Price
Mantel	Delta	1880	5" and 6"	$500.00

Type	Pattern	Circa	Dial	Price
Mantel	Bedford	1880	5"	$350.00

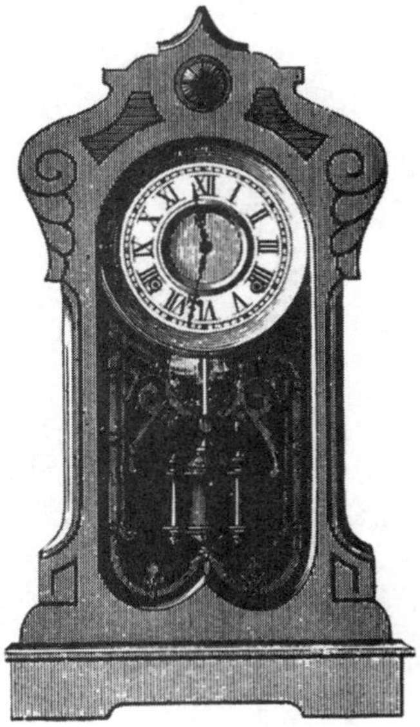

Type	Pattern	Circa	Dial	Price
Mantel	Clifton	1880	6"	$350.00

Type	Pattern	Circa	Dial	Price
Mantel	Rutland	1880	5"	$350.00

Type	Pattern	Circa	Dial	Price
Mantel	Hastings	1880	6"	$375.00

Type	Pattern	Circa	Dial	Price
Mantel	Kenmore	1880	6"	$400.00

Type	Pattern	Circa	Dial	Price
Mantel	Rome	1880	5"	$550.00

Type	Pattern	Circa	Dial	Price
Mantel	Paris	1880	6"	$595.00

Type	Pattern	Circa	Dial	Price
Mantel	Montrose	1880	6"	$450.00

Waterbury

Type	Pattern	Circa	Dial	Price
Mantel	Asia	1880	5"	$850.00

Type	Pattern	Circa	Dial	Price
Mantel	Oxford	1880	6"	$895.00

Type	Pattern	Circa	Dial	Price
Mantel	Lenox	1880	5"	$800.00

Type	Pattern	Circa	Dial	Price
Mantel	China	1880	6"	$950.00

Type	Pattern	Circa	Dial	Price
Hall	No. 78	1918	12"	$10,000.00

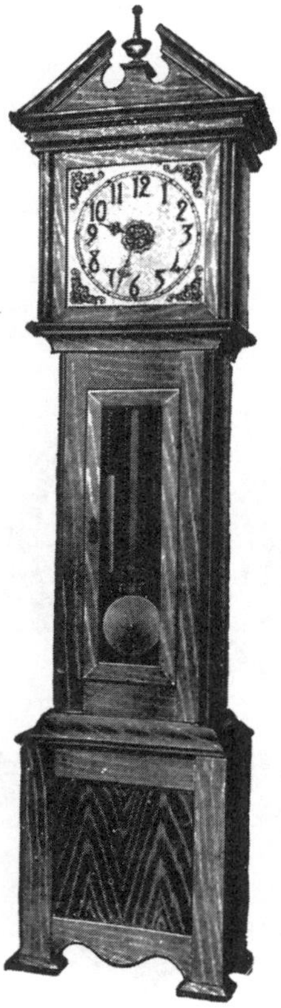

Type	Pattern	Circa	Dial	Price
Hall	No. 74	1918	12"	$10,000.00

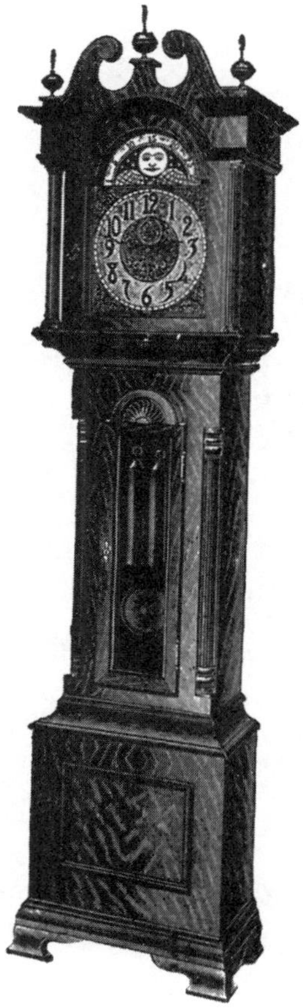

Type	Pattern	Circa	Height	Price
Hall	No. 801	1918	78½"	$10,000.00

THE ANSONIA CLOCK COMPANY.

Type	Pattern	Circa	Dial	Price
Pinion	Bee	1886	2"	$110.00

Type	Pattern	Circa	Dial	Price
Pinion	Bee Alarm	1886	2"	$120.00

Type	Pattern	Circa	Dial	Price
Pinion	Dandy Alarm	1886	3"	$80.00

Type	Pattern	Circa	Dial	Price
Pinion	Dandy	1886	3"	$65.00

Type	Pattern	Circa	Dial	Price
Nickel Novelty	Princess	1886	4"	$65.00

Type	Pattern	Circa	Dial	Price
Nickel Novelty	Planet	1886	4"	$95.00

Type	Pattern	Circa	Dial	Price
Nickel Novelty	Octagon Peep-O'-Day	1886	4"	$60.00

Type	Pattern	Circa	Dial	Price
Nickel Novelty	Peep-O'-Day Strike	1886	4"	$80.00

Type	Pattern	Circa	Dial	Price
Nickel Novelty	Peep-O'-Day	1886	4"	$95.00

Type	Pattern	Circa	Dial	Price
Nickel Novelty	Midget	1886	8"	$80.00

Type	Pattern	Circa	Dial	Price
Nickel Novelty	Octagon Peep-O'-Day Calendar	1886	4"	$100.00

Ansonia

Type	Pattern	Circa	Dial	Price
Nickel Novelty	Standard Time	1886	4"	$80.00

Type	Pattern	Circa	Dial	Price
Nickel Novelty	Octagon	1886	4"	$65.00

Type	Pattern	Circa	Dial	Price
Nickel Novelty	Peep-O'-Day Fancy	1886	6"	$80.00

Type	Pattern	Circa	Dial	Price
Nickel Novelty	Pirate	1886	4"	$65.00

Type	Pattern	Circa	Dial	Price
Nickel Novelty	Peep-O'-Day	1886	6"	$80.00

Type	Pattern	Circa	Dial	Price
Nickel Novelty	Standard Peep-O'-Day	1886	4"	$80.00

Type	Pattern	Circa	Dial	Price
Nickel Novelty	Peep-O'-Day Carriage	1886	3"	$300.00

Type	Pattern	Circa	Dial	Price
Nickel Novelty	Carriage Peep-O'-Day Strike	1886	3"	$300.00

Type	Pattern	Circa	Dial	Price
Nickel Novelty	Peep-O'-Day Carriage-Fancy	1886	3"	$325.00

Type	Pattern	Circa	Dial	Price
Nickel Novelty	Carriage Eight Day	1886	3"	$350.00

Type	Pattern	Circa	Dial	Price
Nickel Novelty	Oriole	1886	3"	$375.00

Type	Pattern	Circa	Dial	Price
Nickel Novelty	Ornamental Carriage	1886	3"	$375.00

Ansonia

Type	Pattern	Circa	Height	Price
Mantel	Small Sharp Gothic	1886	15"	$225.00

Type	Pattern	Circa	Height	Price
Mantel	Cottage	1886	12"	$195.00

Type	Pattern	Circa	Height	Price
Mantel	Neptune	1886	15"	$220.00

Type	Pattern	Circa	Height	Price
Mantel	Cottage, Extra	1886	13"	$200.00

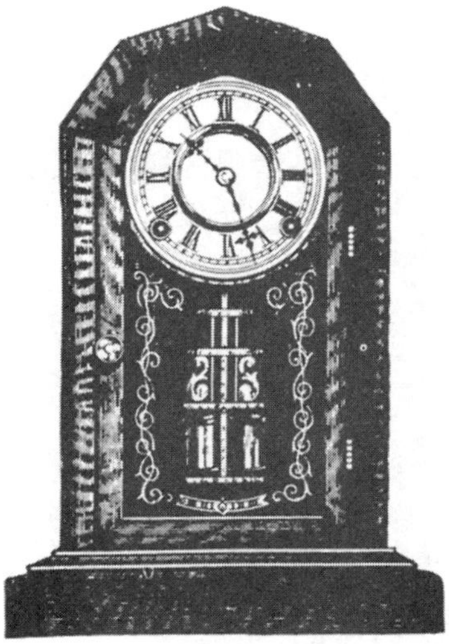

Type	Pattern	Circa	Height	Price
Mantel	Theban	1886	18½"	$275.00

Type	Pattern	Circa	Height	Price
Mantel	O.G. Weight	1886	26"	$250.00

Type	Pattern	Circa	Height	Price
Mantel	Herald	1886	16½"	$245.00

Type	Pattern	Circa	Height	Price
Mantel	Sharp Gothic, V.P.	1886	19½"	$275.00

Type	Pattern	Circa	Height	Price
Mantel	Arcadian	1886	18"	$250.00

Type	Pattern	Circa	Height	Price
Mantel	Standard	1886	18½"	$275.00

Type	Pattern	Circa	Height	Price
Mantel	Tudor, V.P.	1886	19"	$225.00

Type	Pattern	Circa	Height	Price
Mantel	Spartan	1886	18½"	$250.00

Ansonia

Type	Pattern	Circa	Height	Price
Mantel	World	1886	16½"	$375.00

Type	Pattern	Circa	Height	Price
Mantel	Express	1886	21"	$375.00

Type	Pattern	Circa	Height	Price
Mantel	Tribune	1886	18"	$365.00

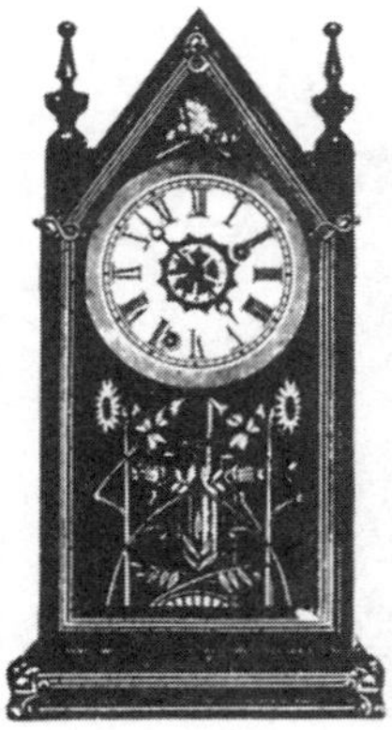

Type	Pattern	Circa	Height	Price
Mantel	Small Decorated Gothic	1886	16"	$275.00

Type	Pattern	Circa	Height	Price
Mantel	Globe	1886	19"	$275.00

Type	Pattern	Circa	Height	Price
Mantel	Decorated Gothic	1886	20"	$275.00

Type	Pattern	Circa	Height	Price
Mantel	Colon	1886	24"	$650.00

Type	Pattern	Circa	Height	Price
Mantel	Greek	1886	17"	$600.00

Type	Pattern	Circa	Height	Price
Mantel	Sun	1886	16"	$575.00

Type	Pattern	Circa	Height	Price
Mantel	Post	1886	20"	$620.00

Type	Pattern	Circa	Height	Price
Mantel	Japan	1886	19"	$550.00

Type	Pattern	Circa	Height	Price
Mantel	Steel	1886	16½"	$550.00

Ansonia

Type	Pattern	Circa	Height	Price
Mantel	Metropolis	1886	24¼"	$650.00

Type	Pattern	Circa	Height	Price
Mantel	Michigan	1886	24¼"	$650.00

Type	Pattern	Circa	Height	Price
Mantel	Jasmine	1886	22½"	$600.00

Type	Pattern	Circa	Height	Price
Mantel	Windsor	1886	21½"	$750.00

Type	Pattern	Circa	Height	Price
Mantel	Triumph	1886	24½"	$750.00

Type	Pattern	Circa	Height	Price
Mantel	Peru	1886	21¼"	$650.00

Type	Pattern	Circa	Height	Price
Mantel	Mobile	1886	20"	$400.00

Type	Pattern	Circa	Height	Price
Mantel	Ranger	1886	20"	$450.00

Type	Pattern	Circa	Height	Price
Mantel	Parisian	1886	23½"	$450.00

Type	Pattern	Circa	Height	Price
Mantel	Tantivy	1886	23"	$400.00

Type	Pattern	Circa	Height	Price
Mantel	Ringgold	1886	24¾"	$375.00

Type	Pattern	Circa	Height	Price
Mantel	Sydney	1886	22"	$400.00

Type	Pattern	Circa	Height	Price
Mantel	Equal	1886	21½"	$375.00

Type	Pattern	Circa	Height	Price
Mantel	Louise	1886	19"	$375.00

Type	Pattern	Circa	Height	Price
Mantel	Lima	1886	24"	$375.00

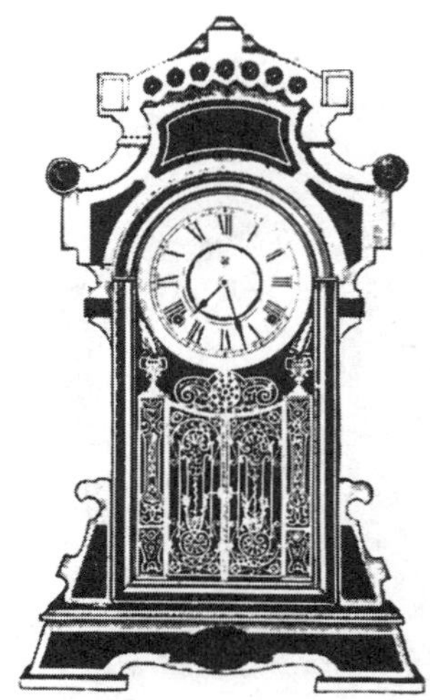

Type	Pattern	Circa	Height	Price
Mantel	Madison	1886	22½"	$375.00

Type	Pattern	Circa	Height	Price
Mantel	Lowell	1886	23½"	$375.00

Type	Pattern	Circa	Height	Price
Mantel	Monarch	1886	24½"	$1,250.00

Type	Pattern	Circa	Height	Price
Mantel	Gallant	1886	22½"	$350.00

Type	Pattern	Circa	Height	Price
Mantel	Ingomar	1886	22"	$375.00

Type	Pattern	Circa	Height	Price
Mantel	Galena	1886	21"	$375.00

Type	Pattern	Circa	Height	Price
Mantel	King	1886	24"	$450.00

Type	Pattern	Circa	Height	Price
Mantel	Cuba	1886	24½"	$400.00

Type	Pattern	Circa	Height	Price
Mantel	Kentucky	1886	22"	$375.00

Ansonia

Type	Pattern	Circa	Height	Price
Mantel	Atlas	1886	21"	$375.00

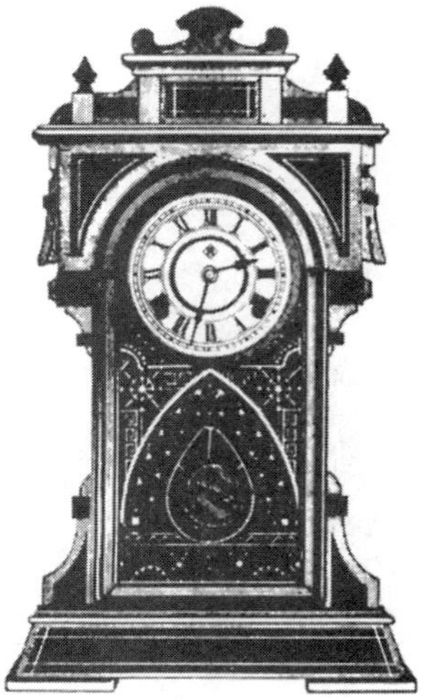

Type	Pattern	Circa	Height	Price
Mantel	Erie	1886	22"	$350.00

Type	Pattern	Circa	Height	Price
Mantel	Alaska	1886	21"	$350.00

Type	Pattern	Circa	Height	Price
Mantel	Farragut	1886	22"	$350.00

Type	Pattern	Circa	Height	Price
Mantel	America	1886	21"	$375.00

Type	Pattern	Circa	Height	Price
Mantel	Fulton	1886	21"	$375.00

Type	Pattern	Circa	Height	Price
Mantel	Canada	1886	21"	$250.00

Type	Pattern	Circa	Height	Price
Mantel	Australia	1886	21"	$250.00

Type	Pattern	Circa	Height	Price
Mantel	Celtic	1886	19"	$250.00

Type	Pattern	Circa	Height	Price
Mantel	Arabia	1886	21"	$250.00

Type	Pattern	Circa	Height	Price
Mantel	Fulda	1886	21"	$250.00

Type	Pattern	Circa	Height	Price
Mantel	Africa	1886	21"	$395.00+

Ansonia

Type	Pattern	Circa	Height	Price
Mantel	Oder	1886	19"	$250.00

Type	Pattern	Circa	Height	Price
Mantel	Aurania	1886	20"	$250.00

Type	Pattern	Circa	Height	Price
Mantel	Oregon	1886	19"	$250.00

Type	Pattern	Circa	Height	Price
Mantel	Baltic	1886	19½"	$250.00

Type	Pattern	Circa	Height	Price
Mantel	Werra	1886	19½"	$250.00

Type	Pattern	Circa	Height	Price
Mantel	Britannic	1886	19½"	$250.00

Type	Pattern	Circa	Height	Price
Mantel	Pizarro & Cortez	1886	20½"	$1,200.00

Type	Pattern	Circa	Height	Price
Mantel	Tasso	1886	20½"	$850.00

Type	Pattern	Circa	Height	Price
Mantel	Fisher & Hunter	1886	21½"	$1,250.00

Type	Pattern	Circa	Height	Price
Mantel	Art & Commerce	1886	20½"	$1,350.00

Type	Pattern	Circa	Height	Price
Mantel	Insult	1886	$21^{3}/_{4}$"	$900.00

Type	Pattern	Circa	Height	Price
Mantel	Provocation	1886	$21^{3}/_{4}$"	$900.00

Type	Pattern	Circa	Height	Price
Mantel	Attila	1886	$21^{1}/_{2}$"	$900.00

Type	Pattern	Circa	Height	Price
Mantel	Mars	1886	$21^{1}/_{2}$"	$900.00

Type	Pattern	Circa	Height	Price
Mantel	Don Juan	1886	22"	$900.00

Type	Pattern	Circa	Height	Price
Mantel	Don Caesar	1886	22"	$900.00

Type	Pattern	Circa	Height	Price
Mantel	Pizarro	1886	21¾"	$900.00

Type	Pattern	Circa	Height	Price
Mantel	Cortez	1886	21¾"	$900.00

Type	Pattern	Circa	Height	Price
Mantel	Denis Papin	1886	16¼"	$700.00

Type	Pattern	Circa	Height	Price
Mantel	Music	1886	21¾"	$900.00

Type	Pattern	Circa	Height	Price
Mantel	Reubens	1886	16¼"	$850.00

Type	Pattern	Circa	Height	Price
Mantel	Combatants	1886	21"	$1,250.00

Type	Pattern	Circa	Height	Price
Mantel	Don Caesar & Don Juan	1886	20½"	$1,250.00

Type	Pattern	Circa	Height	Price
Mantel	Florentine No. 3	1886	13½"	$450.00

Type	Pattern	Circa	Height	Price
Mantel	Florentine No. 6	1886	16"	$350.00

Type	Pattern	Circa	Height	Price
Mantel	Florentine No. 1	1886	15½"	$375.00

Type	Pattern	Circa	Height	Price
Mantel	Antique No. 1	1886	15"	$1,250.00

Type	Pattern	Circa	Height	Price
Mantel	Florentine No. 5	1886	14"	$375.00

Ansonia

Type	Pattern	Circa	Height	Price
Mantel	Parma Figure No. 1069	1886	11"	$650.00

Type	Pattern	Circa	Height	Price
Mantel	Mercury	1886	15"	$750.00

Type	Pattern	Circa	Height	Price
Mantel	Newton	1886	16¼"	$750.00

Type	Pattern	Circa	Height	Price
Mantel	Opera	1886	16¼"	$750.00

Type	Pattern	Circa	Height	Price
Mantel	Troubadour	1886	16¼"	$750.00

Type	Pattern	Circa	Height	Price
Mantel	Industry	1886	16½"	$750.00

Type	Pattern	Circa	Dial	Price
Novelty	Study Ink	1886	4"	$325.00

Type	Pattern	Circa	Dial	Price
Novelty	Lily Ink	1886	4"	$365.00

Type	Pattern	Circa	Dial	Price
Novelty	Gem Ink	1886	4"	$365.00

Type	Pattern	Circa	Dial	Price
Novelty	Butterfly Ink	1886	4"	$395.00

Type	Pattern	Circa	Dial	Price
Novelty	Good Luck	1886	4"	$245.00

Type	Pattern	Circa	Dial	Price
Novelty	Bella	1886	4"	$245.00

Type	Pattern	Circa	Dial	Price
Novelty	Beauty	1886	4"	$195.00

Type	Pattern	Circa	Dial	Price
Novelty	Echo	1886	4"	$195.00

Type	Pattern	Circa	Dial	Price
Novelty	Marguerite	1886	4"	$725.00

Type	Pattern	Circa	Dial	Price
Novelty	Little Dorrit	1886	4"	$550.00

Type	Pattern	Circa	Height	Price
Mantel	Philosopher	1886	15"	$1,000.00

Type	Pattern	Circa	Height	Price
Cabinet	Huntress Swing	1886	25"	$3,000.00

Type	Pattern	Circa	Height	Price
Cabinet	Diana	1886	30"	$4,000.00

Type	Pattern	Circa	Height	Price
Cabinet	Double Figure Swing	1886	27"	$5,000.00

Type	Pattern	Circa	Height	Price
Mantel	Persia	1886	11¼"	$350.00

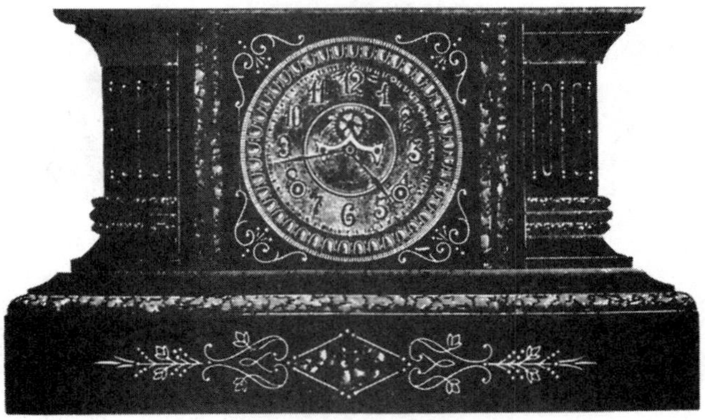

Type	Pattern	Circa	Height	Price
Mantel	Norway	1886	10¼"	$350.00

Type	Pattern	Circa	Height	Price
Mantel	Holland	1886	10¾"	$350.00

Type	Pattern	Circa	Height	Price
Mantel	Como	1886	10"	$350.00

Type	Pattern	Circa	Height	Price
Mantel	Verona	1886	10½"	$500.00

Type	Pattern	Circa	Height	Price
Mantel	Madras	1886	12"	$350.00

Type	Pattern	Circa	Height	Price
Mantel	Sweden	1886	13¼"	$350.00

Type	Pattern	Circa	Height	Price
Mantel	Bombay	1886	14"	$385.00

Type	Pattern	Circa	Height	Price
Mantel	Denmark	1886	11½"	$500.00

Type	Pattern	Circa	Height	Price
Mantel	India	1886	12"	$350.00

Type	Pattern	Circa	Height	Price
Mantel	Calcutta	1886	12"	$395.00

Type	Pattern	Circa	Height	Price
Mantel	Toledo	1886	10¼"	$375.00

Ansonia

Type	Pattern	Circa	Height	Price
Mantel	Nile, with Urn	1886	18"	$600.00

Type	Pattern	Circa	Height	Price
Mantel	Pompeii	1886	10¾"	$650.00

Type	Pattern	Circa	Height	Price
Mantel	Timbrel	1886	13"	$350.00

Type	Pattern	Circa	Height	Price
Mantel	Angelo	1886	12½"	$350.00

Type	Pattern	Circa	Height	Price
Mantel	Timbrel No. 1	1886	12¼"	$325.00

Type	Pattern	Circa	Height	Price
Mantel	Timbrel No. 2	1886	10½"	$325.00

Type	Pattern	Circa	Height	Price
Mantel	La Duchesse	1886	10½"	$200.00

Type	Pattern	Circa	Height	Price
Mantel	Madrid	1886	10¼"	$200.00

Type	Pattern	Circa	Height	Price
Mantel	Italy	1886	11¼"	$250.00

Type	Pattern	Circa	Height	Price
Mantel	LaFrance, with Urn	1886	16¾"	$375.00

Type	Pattern	Circa	Height	Price
Mantel	Palermo	1886	10¼"	$250.00

Type	Pattern	Circa	Height	Price
Mantel	Rosalind	1886	12¼"	$500.00

Ansonia

Type	Pattern	Circa	Height	Price
Mantel	St. Clair	1886	11¼"	$300.00

Type	Pattern	Circa	Height	Price
Mantel	Messina	1886	11"	$275.00

Type	Pattern	Circa	Height	Price
Mantel	Servia	1886	10½"	$275.00

Type	Pattern	Circa	Height	Price
Mantel	Euclid with Figure No. 1042	1886	16"	$475.00

Type	Pattern	Circa	Height	Price
Mantel	Nero	1886	12¾"	$250.00

Type	Pattern	Circa	Height	Price
Mantel	Spain	1886	13"	$250.00

Type	Pattern	Circa	Height	Price
Cabinet	Turin	1886	16½"	$350.00

Type	Pattern	Circa	Height	Price
Cabinet	Cairo	1886	15½"	$375.00

Type	Pattern	Circa	Height	Price
Cabinet	Rome	1886	15"	$350.00

Type	Pattern	Circa	Height	Price
Mantel	Egypt	1886	10¾"	$350.00

Type	Pattern	Circa	Height	Price
Mantel	Russia	1886	11"	$375.00

Type	Pattern	Circa	Height	Price
Cabinet	Tivoli	1886	15"	$275.00

Type	Pattern	Circa	Height	Price
Cabinet	Tunis	1886	14"	$275.00

Type	Pattern	Circa	Height	Price
Cabinet	Milan	1886	13"	$225.00

Type	Pattern	Circa	Height	Price
Cabinet	Unique	1886	10½"	$185.00

Type	Pattern	Circa	Height	Price
Cabinet	Imogene	1886	10½"	$245.00

Type	Pattern	Circa	Height	Price
Mantel	Lisbon	1886	10½"	$225.00

Type	Pattern	Circa	Height	Price
Mantel	Florentine No. 2	1886	14"	$2,250.00

Type	Pattern	Circa	Height	Price
Mantel	Florentine No. 4	1886	15"	$2,500.00

Type	Pattern	Circa	Height	Price
Mantel	Etruscan	1886	14½"	$1,550.00

Type	Pattern	Circa	Height	Price
Mantel	Regent	1886	20½"	$6,500.00

Type	Pattern	Circa	Height	Price
Mantel	Renaissance	1886	15"	$3,500.00

Type	Pattern	Circa	Height	Price
Novelty	Symbol, Extra	1886	15¼"	$675.00

Type	Pattern	Circa	Height	Price
Novelty	China	1886	12"	$450.00

Type	Pattern	Circa	Height	Price
Novelty	Richelieu	1886	18½"	$650.00

Type	Pattern	Circa	Height	Price
Novelty	Cabinet Antique	1886	20"	$725.00

Type	Pattern	Circa	Dial	Price
Lever	Ansonia	1886	4"	$265.00

Type	Pattern	Circa	Dial	Price
Lever	Octagon R.C.	1886	6"	$265.00

Type	Pattern	Circa	Dial	Price
Lever	Cable	1886	4"	$295.00

Type	Pattern	Circa	Dial	Price
Lever	Ansonia	1886	12"	$295.00

Type	Pattern	Circa	Dial	Price
Lever	Brass or Nickel	1886	4"	$265.00

Ansonia

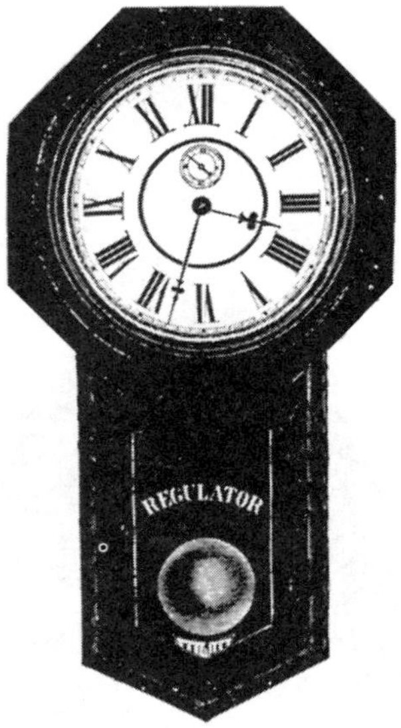

Type	Pattern	Circa	Dial	Price
Wall	Office Regulator	1886	12"	$500.00

Type	Pattern	Circa	Dial	Price
Wall	Regulator A, Calendar	1886	12"	$650.00

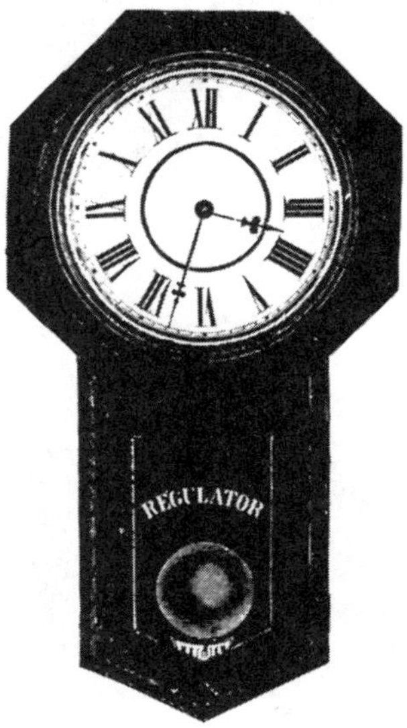

Type	Pattern	Circa	Dial	Price
Wall	Regulator A	1886	12"	$650.00

Type	Pattern	Circa	Dial	Price
Wall	English Drop	1886	12"	$450.00

Type	Pattern	Circa	Dial	Price
Wall	Office No. 2	1886	12"	$350.00

Type	Pattern	Circa	Dial	Price
Wall	Kobe	1886	10"	$625.00

Type	Pattern	Circa	Dial	Price
Wall	Drop Octagon	1886	10"	$425.00

Type	Pattern	Circa	Dial	Price
Wall	Reflector	1886	6"	$500.00

Type	Pattern	Circa	Dial	Price
Wall	Drop Octagon	1886	10"	$400.00

Type	Pattern	Circa	Dial	Price
Wall	Drop Octagon, R.C.	1886	12"	$400.00

Type	Pattern	Circa	Dial	Price
Wall	Queen Elizabeth	1886	8"	$950.00

Ansonia

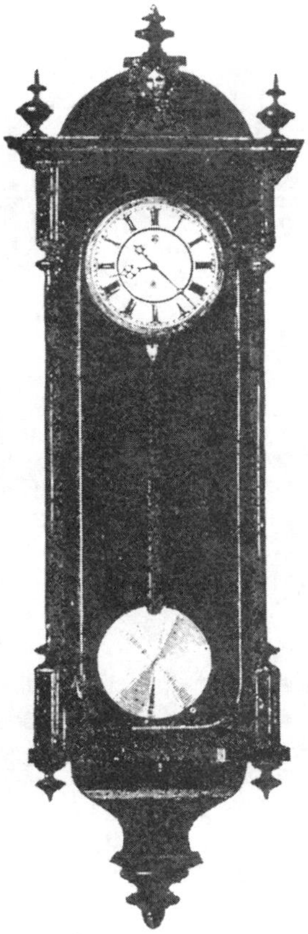

Type	Pattern	Circa	Dial	Price
Wall	Capitol	1886	8"	$1,250.00

Type	Pattern	Circa	Dial	Price
Weight	Mecca	1886	8"	$1,250.00

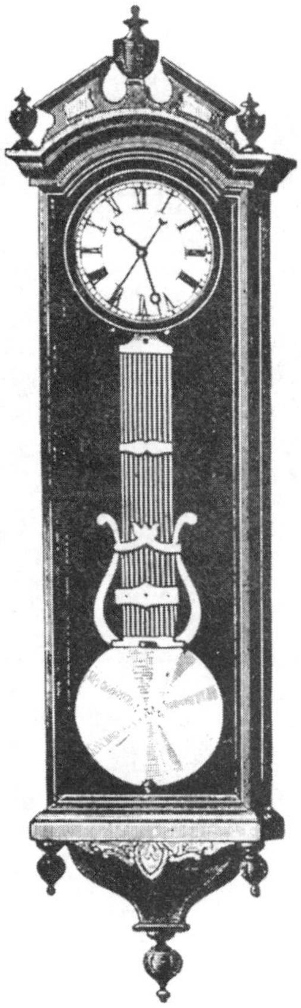

Type	Pattern	Circa	Dial	Price
Weight	Regulator No. 4	1886	12"	$1,450.00

Type	Pattern	Circa	Dial	Price
Weight	Medina	1886	8"	$1,250.00

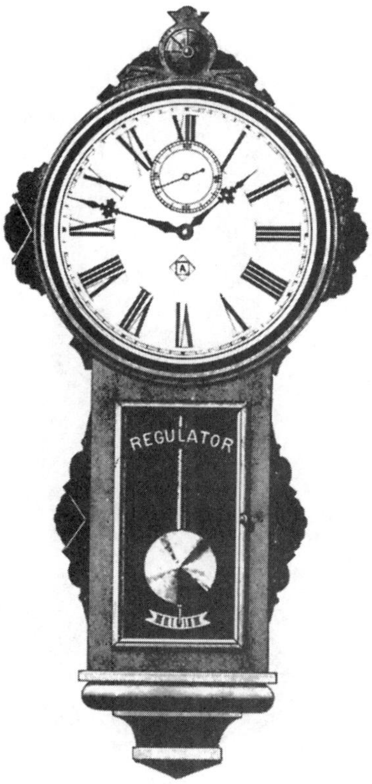

Type	Pattern	Circa	Dial	Price
Weight	General	1886	18"	$1,250.00

Type	Pattern	Circa	Dial	Price
Weight	Regulator No. 14	1886	8"	$1,500.00

Type	Pattern	Circa	Dial	Price
Wall	Dispatch	1886	12"	$650.00

Type	Pattern	Circa	Dial	Price
Weight	Antique Hanging Clock	1886	9½"	$3,500.00

Ansonia

Type	Pattern	Circa	Dial	Price
Wall	Santa Fe	1886	10"	$1,500.00

Type	Pattern	Circa	Dial	Price
Wall	Forrest	1886	8"	$1,500.00

Type	Pattern	Circa	Dial	Price
Wall	Prompt	1886	8"	$1,250.00

Type	Pattern	Circa	Dial	Price
Wall	Bagdad	1886	8"	$1,500.00

Type	Pattern	Circa	Dial	Price
Weight	Antique Hall Clock	1886	10"	$12,500.00

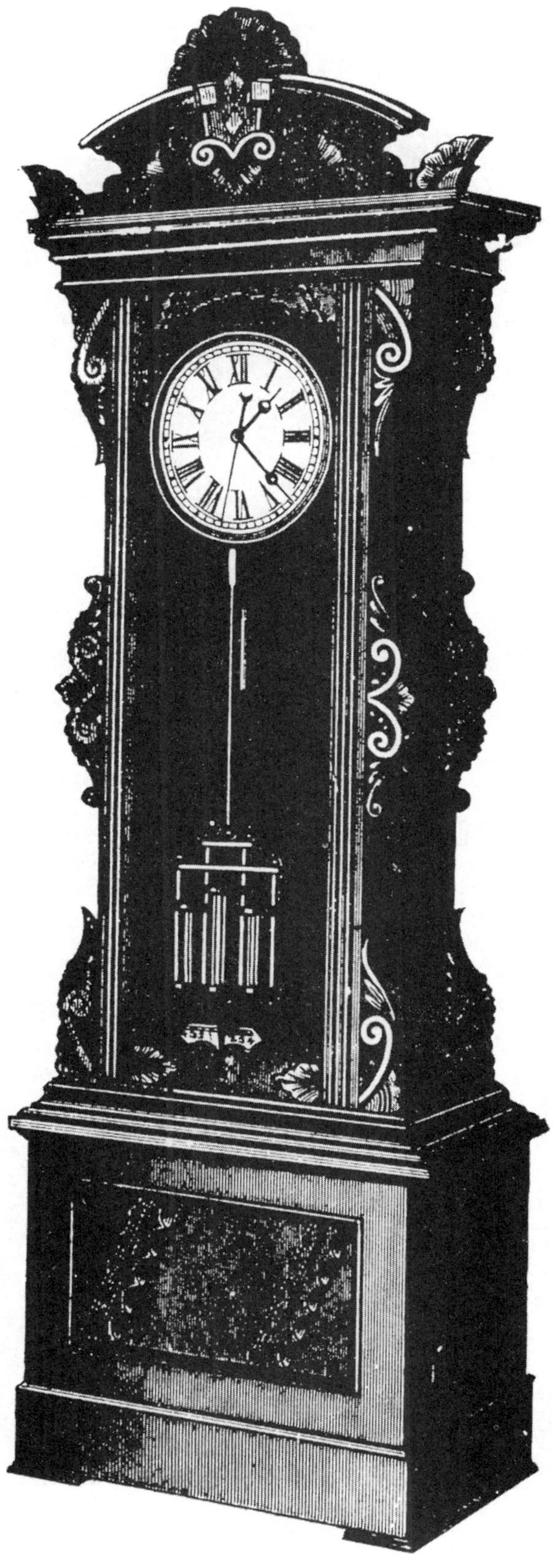

Type	Pattern	Circa	Dial	Price
Weight	New Regulator	1886	12"	$15,000.00

Type	Pattern	Circa	Dial	Price
Weight	Regulator No. 8	1886	16"	$15,000.00

Type	Pattern	Circa	Dial	Price
Weight	Regulator No. 11	1886	14"	$15,000.00

Type	Pattern	Circa	Dial	Price
Dresser	Hazel	1930	2½"	$45.00

Type	Pattern	Circa	Height	Price
Dresser	Peerless	1930	7¼"	$40.00

Type	Pattern	Circa	Height	Price
Artlarm	Easel	1930	4¾"	$45.00

Type	Pattern	Circa	Dial	Price
Dresser	Quality	1930	3¾"	$25.00

Type	Pattern	Circa	Height	Price
Artlarm	Square	1930	4½"	$25.00

Type	Pattern	Circa	Height	Price
Artlarm	Abbey	1930	5¼"	$25.00

New Haven

Type	Pattern	Circa	Dial	Price
Cabinet	Malo	1918	5"	$80.00

Type	Pattern	Circa	Dial	Price
Cabinet	Myra	1918	5"	$80.00

Type	Pattern	Circa	Dial	Price
Cabinet	Colonial	1918	7"	$65.00

Type	Pattern	Circa	Dial	Price
Cabinet	Adam No. 2	1918	7"	$65.00

Type	Pattern	Circa	Dial	Price
Cabinet	Jacobean No. 1	1918	7"	$65.00

Type	Pattern	Circa	Dial	Price
Cabinet	Stuart	1918	7"	$65.00

Type	Pattern	Circa	Dial	Price
Wall	Madrid	1918	12½"	$350.00

Type	Pattern	Circa	Dial	Price
Wall	Mission Hanging	1918	12"	$375.00

Type	Pattern	Circa	Dial	Price
Novelty	Nathalia	1918	2"	$100.00

Type	Pattern	Circa	Dial	Price
Novelty	Eberle	1918	2"	$135.00

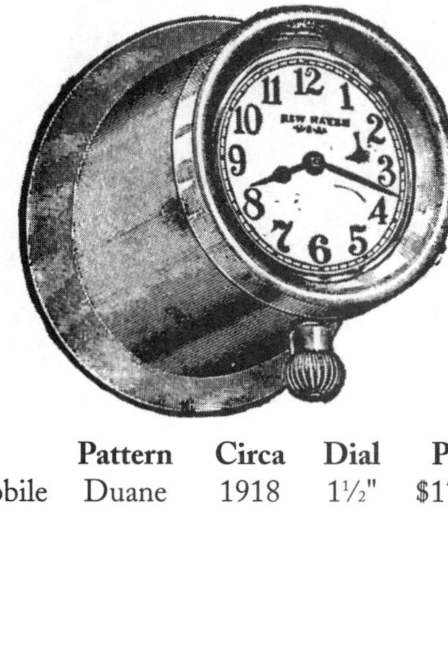

Type	Pattern	Circa	Dial	Price
Automobile	Duane	1918	1½"	$175.00

Type	Pattern	Circa	Dial	Price
Novelty	Harbin	1918	2¼"	$65.00

Type	Pattern	Circa	Dial	Price
Novelty	Ionia	1918	2"	$65.00

Type	Pattern	Circa	Dial	Price
Novelty	Ivy	1918	2"	$65.00

Type	Pattern	Circa	Dial	Price
Novelty	Haines	1918	2¼"	$110.00

Type	Pattern	Circa	Dial	Price
Novelty	Hardy	1918	2¼"	$65.00

Type	Pattern	Circa	Dial	Price
Mantel	Bingen	1918	6"	$150.00

Type	Pattern	Circa	Dial	Price
Mantel	Ottawa	1918	6"	$165.00

Type	Pattern	Circa	Dial	Price
Mantel	No. 3452	1918	6"	$195.00

Type	Pattern	Circa	Dial	Price
Mantel	No. 3451	1918	6"	$165.00

Type	Pattern	Circa	Dial	Price
Mantel	Loring	1918	6"	$165.00

Type	Pattern	Circa	Dial	Price
Mantel	Woodbridge	1918	6"	$165.00

New Haven

Type	Pattern	Circa	Dial	Price
Mantel	Thurston	1918	4"	$200.00

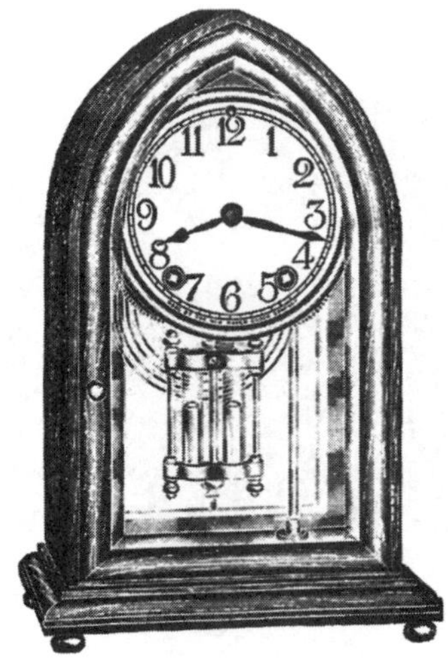

Type	Pattern	Circa	Dial	Price
Mantel	Theone	1918	4"	$300.00

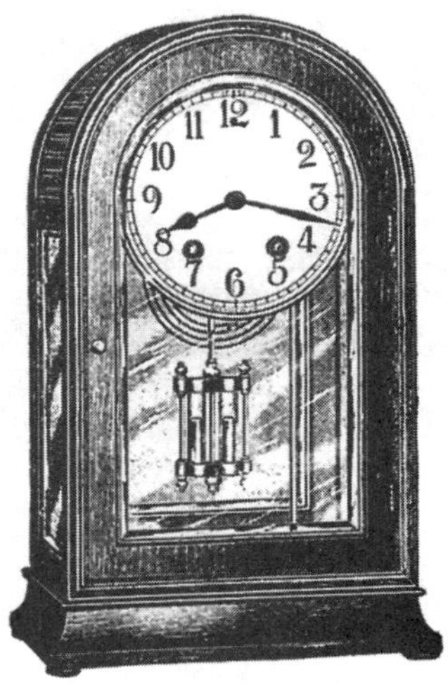

Type	Pattern	Circa	Dial	Price
Mantel	Thermo	1918	7"	$250.00

Type	Pattern	Circa	Dial	Price
Mantel	Taunton	1918	7"	$425.00

Type	Pattern	Circa	Dial	Price
Mantel	Thorpe	1918	6"	$250.00

Type	Pattern	Circa	Dial	Price
Mantel	Theta	1918	7"	$250.00

Type	Pattern	Circa	Dial	Price
Wall	Mimosa	1918	8"	$950.00

Type	Pattern	Circa	Dial	Price
Wall	Marino	1918	8"	$950.00

Type	Pattern	Circa	Dial	Price
Wall	Mirada	1918	8"	$900.00

Type	Pattern	Circa	Dial	Price
Wall	Riverton	1918	12"	$975.00

Type	Pattern	Circa	Dial	Price
Wall	Madero	1918	8"	$1,200.00

New Haven

Type	Pattern	Circa	Dial	Price
Mantel	Mersey Line No. 2	1918	6"	$250.00

Type	Pattern	Circa	Dial	Price
Mantel	Merchants' Line No. 1	1918	6"	$350.00

Type	Pattern	Circa	Dial	Price
Mantel	Mersey Line No. 1	1918	6"	$250.00

Type	Pattern	Circa	Dial	Price
Mantel	Merchants' Line No. 2	1918	6"	$350.00

Type	Pattern	Circa	Dial	Price
Mantel	Mersey Line No. 3	1918	6"	$250.00

Type	Pattern	Circa	Dial	Price
Mantel	Merchant's Line No. 3	1918	6"	$350.00

Type	Pattern	Circa	Dial	Price
Wall	Grecian	1918	10"	$1,395.00

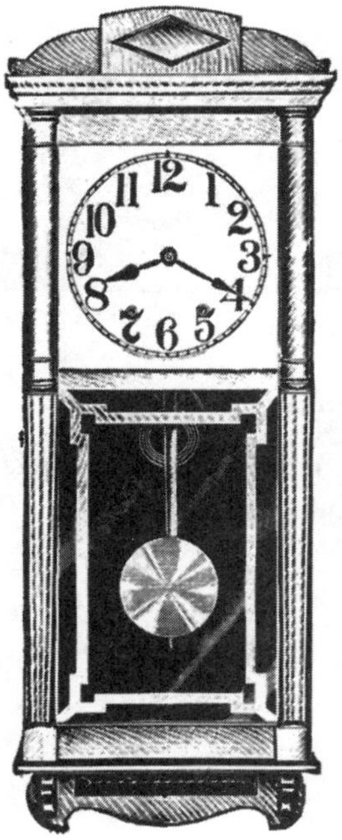

Type	Pattern	Circa	Dial	Price
Wall	Marne	1918	10"	$1,200.00

Type	Pattern	Circa	Dial	Price
Wall	Saturn	1918	12"	$1,500.00

Type	Pattern	Circa	Dial	Price
Wall	Stanton	1918	12"	$1,650.00

Type	Pattern	Circa	Dial	Price
Mantel	Oval No. 53	1918	3½"	$75.00

Type	Pattern	Circa	Dial	Price
Mantel	Queda	1918	3½"	$75.00

Type	Pattern	Circa	Dial	Price
Mantel	Quincy	1918	3½"	$75.00

Type	Pattern	Circa	Dial	Price
Mantel	Quinby	1918	3½"	$75.00

Type	Pattern	Circa	Dial	Price
Mantel	Quayle	1918	3½"	$75.00

Type	Pattern	Circa	Dial	Price
Cabinet	Tambour No. 8	1918	5"	$75.00

Type	Pattern	Circa	Dial	Price
Cabinet	Olympus	1918	6"	$90.00

Type	Pattern	Circa	Dial	Price
Cabinet	Tambour No. 18	1918	6"	$80.00

Type	Pattern	Circa	Dial	Price
Cabinet	Tambour No. 1	1918	7"	$110.00

Type	Pattern	Circa	Dial	Price
Cabinet	Tambour No. 14	1918	6"	$75.00

Type	Pattern	Circa	Dial	Price
Cabinet	Tambour No. 6	1918	7"	$75.00

Type	Pattern	Circa	Dial	Price
Cabinet	Tambour No. 9	1918	7½"	$80.00

New Haven

Type	Pattern	Circa	Dial	Price
Novelty	Harmon	1918	2¼"	$80.00

Type	Pattern	Circa	Dial	Price
Novelty	Boyce	1918	2"	$80.00

Type	Pattern	Circa	Dial	Price
Novelty	Harvey	1918	2¼"	$80.00

Type	Pattern	Circa	Dial	Price
Novelty	Ives	1918	2"	$75.00

Type	Pattern	Circa	Dial	Price
Novelty	Hodges	1918	2¼"	$75.00

Type	Pattern	Circa	Dial	Price
Novelty	Halstead	1918	2¼"	$100.00

Type	Pattern	Circa	Dial	Price
Mantel	Style B Merchants Line	1918	6"	$350.00

Type	Pattern	Circa	Dial	Price
Mantel	Camden	1918	6"	$350.00

Type	Pattern	Circa	Dial	Price
Mantel	Sharp Gothic	1918	6"	$450.00

Type	Pattern	Circa	Dial	Price
Mantel	Style B Gypsy Line	1918	6"	$225.00

Type	Pattern	Circa	Dial	Price
Mantel	San Martin Dandy "Den Clock"	1918	6"	$250.00

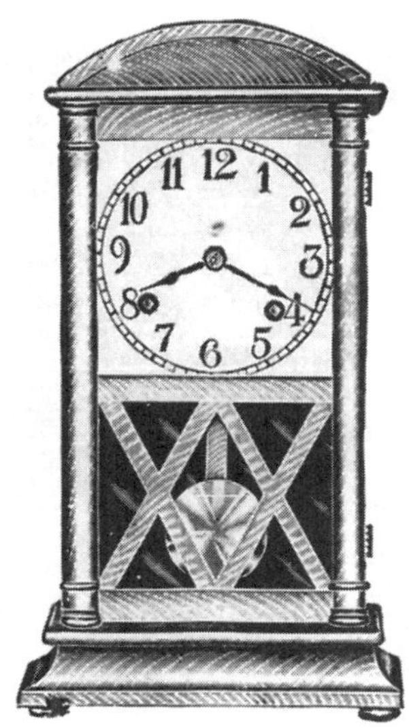

Type	Pattern	Circa	Dial	Price
Mantel	Inwood	1918	5"	$265.00

New Haven

Type	Pattern	Circa	Dial	Price
Wall	Erie	1918	12"	$395.00

Type	Pattern	Circa	Dial	Price
Wall	Referee	1918	12"	$450.00

Type	Pattern	Circa	Dial	Price
Wall	Drop Octagon	1918	8"	$300.00

Type	Pattern	Circa	Dial	Price
Wall	Drop Octagon	1918	10"	$300.00

Type	Pattern	Circa	Dial	Price
Wall	Saxon	1918	10"	$350.00

Type	Pattern	Circa	Dial	Price
Artlarm	Tick-Tock	1930	3¾"	$45.00

Type	Pattern	Circa	Dial	Price
Artlarm	Tat-Too	1930	4¼"	$45.00

Type	Pattern	Circa	Dial	Price
Artlarm	Tat-Too Jr.	1930	2¼"	$45.00

Type	Pattern	Circa	Height	Price
Artlarm	Abbey, Jr.	1930	3½"	$25.00

Type	Pattern	Circa	Dial	Price
Artlarm	Tiny Tot	1930	2¼"	$45.00

Type	Pattern	Circa	Height	Price
Artlarm	Doris	1930	3½"	$30.00

Type	Pattern	Circa	Dial	Price
Wall	Foyer	1930	12"	$90.00

Type	Pattern	Circa	Dial	Price
Wall	Kitchenette	1930	5½"	$75.00

Type	Pattern	Circa	Dial	Price
Alarm	Columbia	1930	4½"	$40.00

Type	Pattern	Circa	Dial	Price
Wall	Clinton	1930	12"	$275.00

Type	Pattern	Circa	Dial	Price
Wall	Office	1930	10"	$275.00

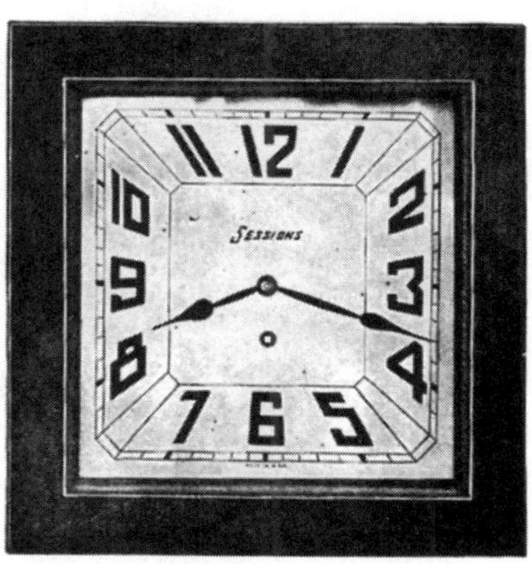

Type	Pattern	Circa	Dial	Price
Wall	Gallery	1930	12"	$200.00

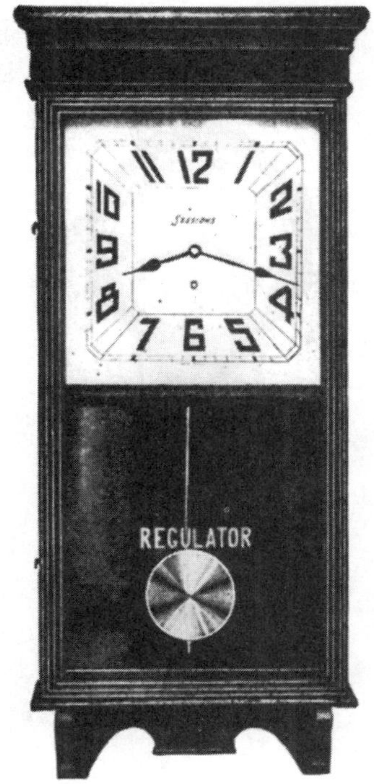

Type	Pattern	Circa	Dial	Price
Wall	Regulator No. 2	1930	12"	$275.00

Type	Pattern	Circa	Dial	Price
Wall	Regulator No. 3	1930	12"	$650.00

Type	Pattern	Circa	Dial	Price
Banjo	Narragansett	1930	5"	$150.00

Type	Pattern	Circa	Dial	Price
Banjo	Lexington	1930	6"	$150.00

Type	Pattern	Circa	Dial	Price
Banjo	Provincetown	1930	5"	$150.00

Type	Pattern	Circa	Dial	Price
Banjo	York	1930	6"	$150.00

Type	Pattern	Circa	Dial	Price
Banjo	Revere	1930	6"	$325.00

Type	Pattern	Circa	Dial	Price
Banjo	Halifax	1930	6"	$325.00

Type	Pattern	Circa	Dial	Price
Banjo	Salem	1930	8"	$350.00

Type	Pattern	Circa	Dial	Price
Mantel	Goldenrod	1930	5"	$160.00

Type	Pattern	Circa	Dial	Price
Mantel	Melrose	1930	5"	$160.00

Type	Pattern	Circa	Dial	Price
Mantel	No. 639	1930	5"	$180.00

Type	Pattern	Circa	Dial	Price
Mantel	Marigold	1930	5"	$180.00

Type	Pattern	Circa	Dial	Price
Mantel	Goldstar	1930	5"	$180.00

Type	Pattern	Circa	Dial	Price
Mantel	Arcadia	1930	5"	$180.00

Type	Pattern	Circa	Dial	Price
Mantel	Black Marbleized, No. 1	1930	5"	$200.00

Type	Pattern	Circa	Dial	Price
Mantel	No. 640	1930	5"	$180.00

Type	Pattern	Circa	Dial	Price
Mantel	Black Marbleized, No. 2	1930	5"	$180.00

Type	Pattern	Circa	Dial	Price
Mantel	Ardmore	1930	5"	$200.00

Type	Pattern	Circa	Dial	Price
Mantel	Black Marbleized, No. 3	1930	5"	$180.00

Type	Pattern	Circa	Dial	Price
Mantel	Baldwin	1930	5"	$200.00

Sessions

Type	Pattern	Circa	Height	Price
Mantel	Mozart	1918	10½"	$200.00

Type	Pattern	Circa	Height	Price
Mantel	Celtic	1918	10½"	$200.00

Type	Pattern	Circa	Height	Price
Mantel	Manhattan	1918	11"	$200.00

Type	Pattern	Circa	Dial	Price
Mantel	Corinth	1918	5"	$180.00

Type	Pattern	Circa	Height	Price
Mantel	Elton	1918	10¼"	$160.00

Type	Pattern	Circa	Height	Price
Mantel	Ideal	1918	10½"	$160.00

Type	Pattern	Circa	Dial	Price
Mantel	Inlay No. 4	1930	5"	$60.00

Type	Pattern	Circa	Dial	Price
Mantel	Berkeley	1930	6"	$60.00

Type	Pattern	Circa	Dial	Price
Mantel	Bellair	1930	6"	$60.00

Type	Pattern	Circa	Dial	Price
Mantel	Inlay No. 6	1930	5"	$60.00

Type	Pattern	Circa	Dial	Price
Mantel	Inlay No. 5	1930	5"	$60.00

Type	Pattern	Circa	Dial	Price
Mantel	Beverly	1930	6"	$60.00

Sessions

Type	Pattern	Circa	Dial	Price
Mantel	Duet No. 1	1930	6"	$60.00

Type	Pattern	Circa	Dial	Price
Mantel	Dulciana	1930	6"	$60.00

Type	Pattern	Circa	Dial	Price
Mantel	Dulce	1930	6"	$60.00

Type	Pattern	Circa	Dial	Price
Mantel	Duet No. 3	1930	6"	$60.00

Type	Pattern	Circa	Dial	Price
Mantel	Duet No. 2	1930	6"	$60.00

Type	Pattern	Circa	Dial	Price
Mantel	Dundee	1930	6"	$60.00

Type	Pattern	Circa	Dial	Price
Mantel	Westminster No. 1	1930	6"	$160.00

Type	Pattern	Circa	Dial	Price
Mantel	Westminster B	1930	6"	$160.00

Type	Pattern	Circa	Dial	Price
Mantel	Westminster A	1930	6"	$160.00

Type	Pattern	Circa	Dial	Price
Mantel	Westminster No. 3	1930	6"	$160.00

Type	Pattern	Circa	Dial	Price
Mantel	Westminster No. 2	1930	6"	$175.00

Type	Pattern	Circa	Dial	Price
Mantel	Westminster C	1930	6"	$175.00

Sessions

Type	Pattern	Circa	Dial	Price
Mantel	Cottage	1930	5"	$195.00

Type	Pattern	Circa	Dial	Price
Mantel	Leader	1930	5½"	$120.00

Type	Pattern	Circa	Dial	Price
Mantel	Cottage	1930	5"	$195.00

Type	Pattern	Circa	Dial	Price
Mantel	Leader	1930	5½"	$120.00

Type	Pattern	Circa	Dial	Price
Mantel	Cottage	1930	5"	$195.00

Type	Pattern	Circa	Dial	Price
Mantel	Leader	1930	5½"	$120.00

Type	Pattern	Circa	Dial	Price
Mantel	Montreal	1930	6"	$150.00

Type	Pattern	Circa	Dial	Price
Mantel	Winlock	1930	6"	$150.00

Type	Pattern	Circa	Dial	Price
Mantel	Wendell	1930	6"	$150.00

Type	Pattern	Circa	Dial	Price
Semi-miniature Mantel	Druid	1930	2½"	$110.00

Type	Pattern	Circa	Dial	Price
Mantel	Windsor	1930	6"	$150.00

Type	Pattern	Circa	Dial	Price
Semi-miniature Mantel	Dawn	1930	2½"	$110.00

Type	Pattern	Circa	Dial	Price
Mantel	Walton	1930	6"	$150.00

Type	Pattern	Circa	Dial	Price
Semi-miniature Mantel	Dart	1930	2½"	$110.00

Type	Pattern	Circa	Height	Price
Mantel	Concord	1930	13"	$95.00

Type	Pattern	Circa	Dial	Price
Semi-miniature Mantel	Easel No. 4	1930	2½"	$110.00

Type	Pattern	Circa	Dial	Price
Semi-miniature Mantel	Secretary No. 1	1930	4½"	$110.00

Type	Pattern	Circa	Dial	Price
Semi-miniature Mantel	Easel No. 3	1930	2½"	$110.00

Type	Pattern	Circa	Dial	Price
Hall	No. 9002	1930	12"	$7,500.00

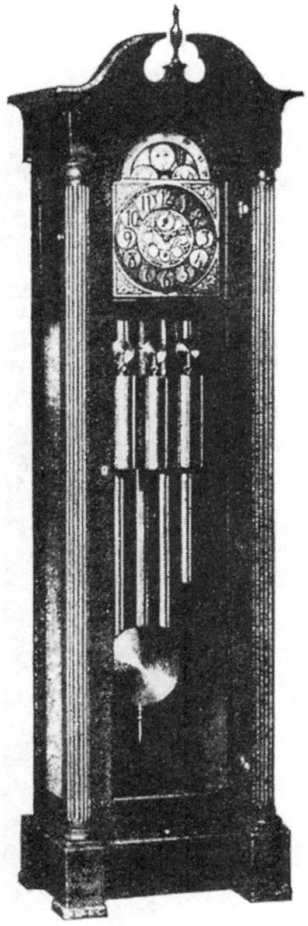

Type	Pattern	Circa	Height	Price
Hall	No. 9003	1930	6' 6½"	$7,500.00

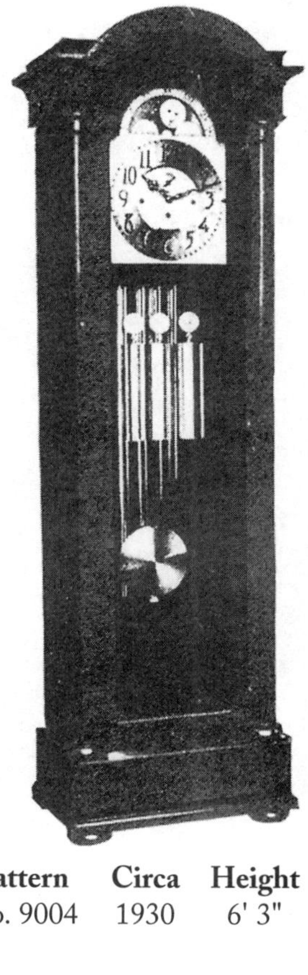

Type	Pattern	Circa	Height	Price
Hall	No. 9004	1930	6' 3"	$7,500.00

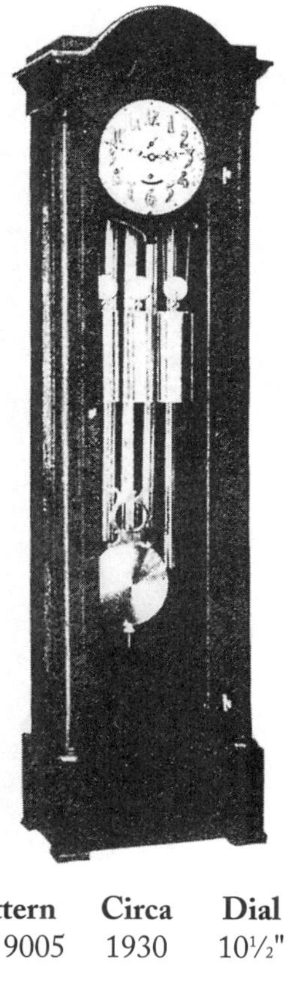

Type	Pattern	Circa	Dial	Price
Hall	No. 9005	1930	10½"	$7,500.00

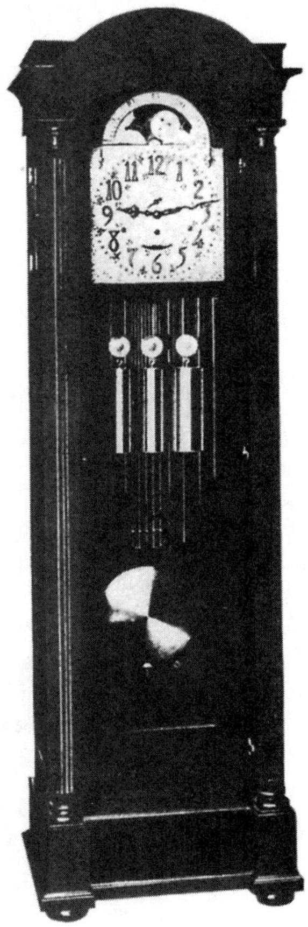

Type	Pattern	Circa	Height	Price
Hall	No. 9000	1930	7' 1"	$7,500.00

Type	Pattern	Circa	Dial	Price
Hall	No. 9001	1930	7' 6"	$7,500.00

THE E. HOWARD CLOCK COMPANY

ESTABLISHED 1842

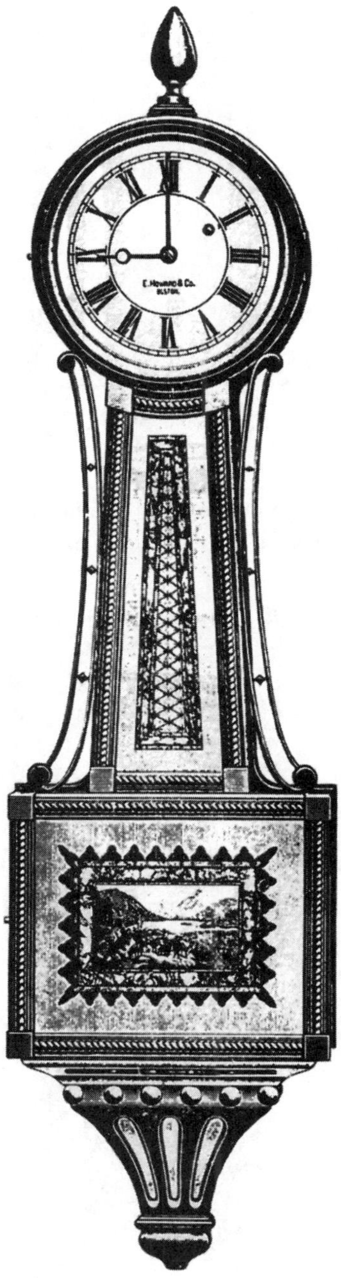

Type	Pattern	Circa	Dial	Price
Wall	Colonial	1880	6"	$9,000.00

Type	Pattern	Circa	Dial	Price
Wall	No. 36 Regulator	1880	14"	$2,400.00

Type	Pattern	Circa	Dial	Price
Wall	No. 11	1880	11"	$6,000.00

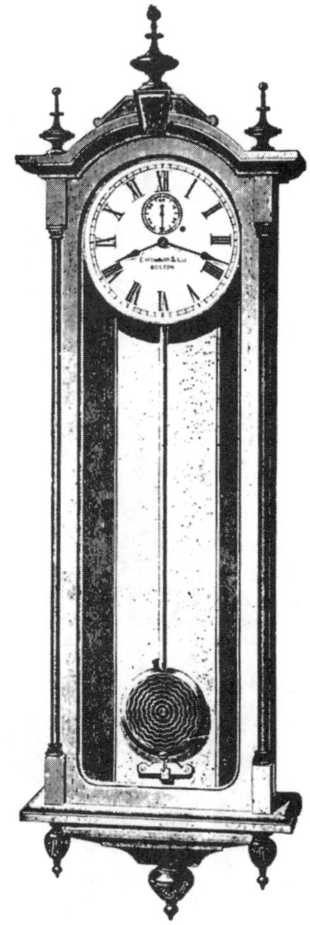

Type	Pattern	Circa	Dial	Price
Wall	No. 71 Regulator	1880	12"	$3,000.00

Type	Pattern	Circa	Dial	Price
Wall	No. 58	1880	12"	$1,800.00

Type	Pattern	Circa	Dial	Price
Wall	No. 85 Regulator	1880	14"	$1,750.00

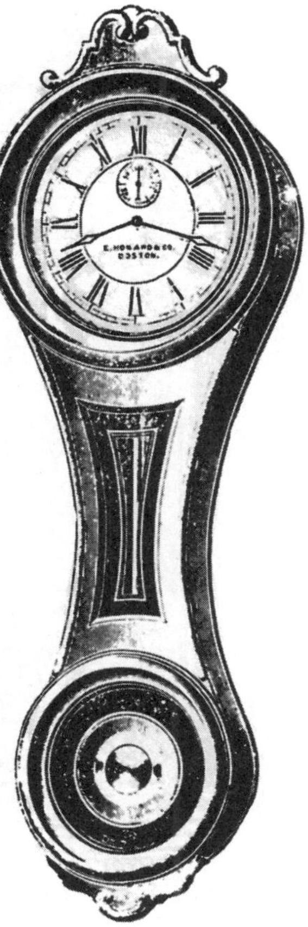

Type	Pattern	Circa	Dial	Price
Wall	No. 6 Regulator	1880	14"	$2,550.00

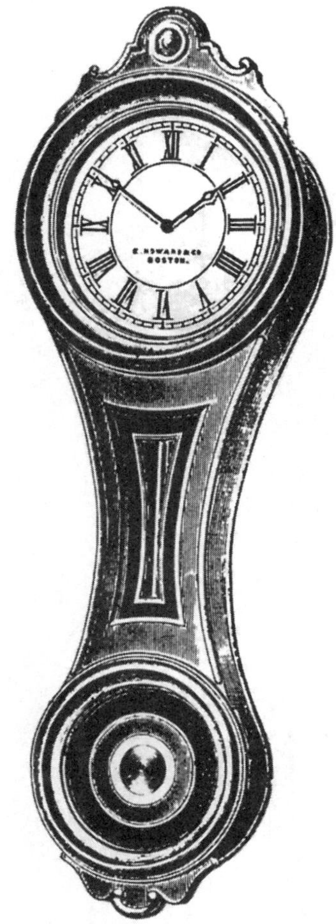

Type	Pattern	Circa	Dial	Price
Wall	No. 7	1880	8" – 12"	$2,400.00

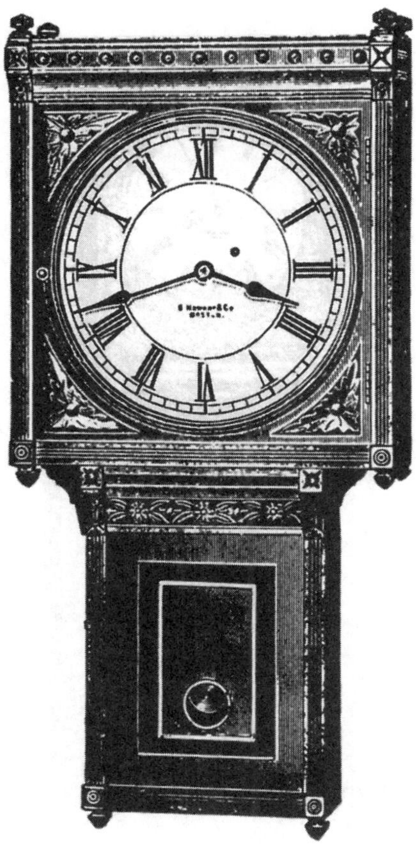

Type	Pattern	Circa	Dial	Price
Wall	No. 75	1880	12"	$1,750.00

Type	Pattern	Circa	Dial	Price
Wall	No. 70	1880	12"	$1,200.00

Howard

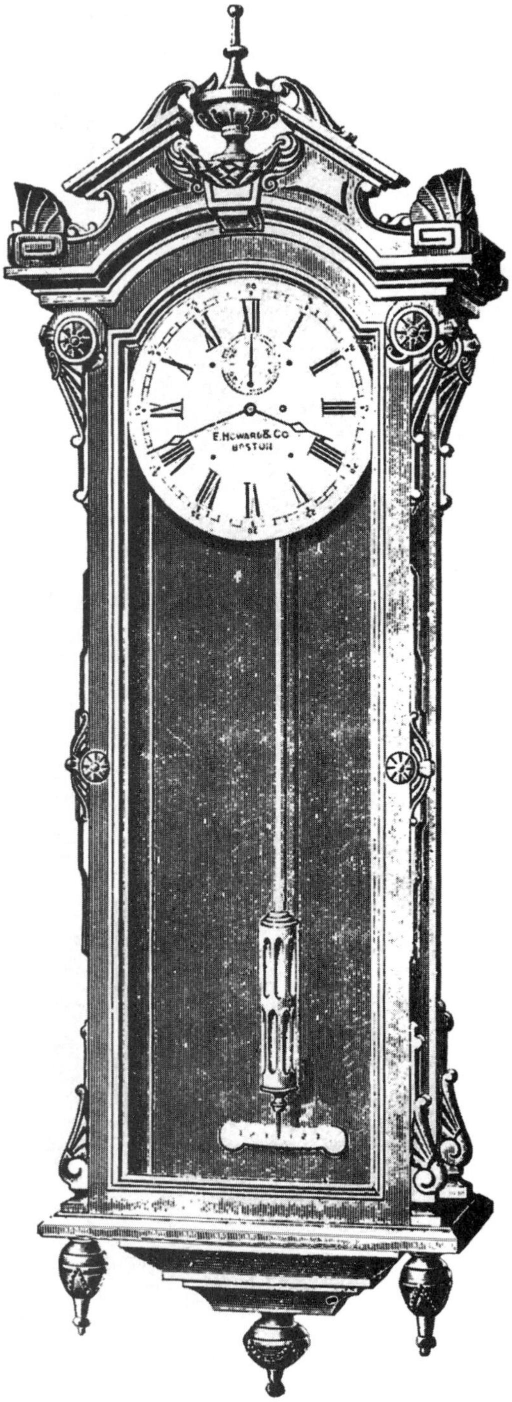

Type	Pattern	Circa	Dial	Price
Wall	No. 60 Regulator	1880	14"	$12,000.00

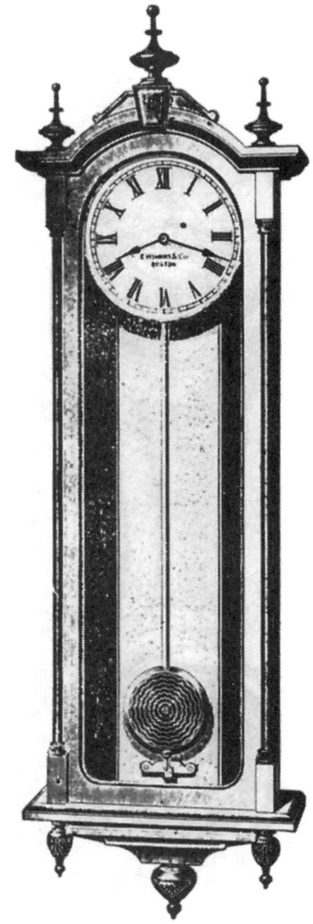

Type	Pattern	Circa	Dial	Price
Wall	No. 59	1880	10"	$1,950.00

Type	Pattern	Circa	Dial	Price
Wall	No. 75 A	1880	14"	$2,000.00

Type	Pattern	Circa	Dial	Price
Wall	No. 86 Regulator	1880	14"	$6,000.00

Type	Pattern	Circa	Dial	Price
Wall	No. 86-A Regulator	1880	12"	$2,400.00

Type	Pattern	Circa	Dial	Price
Wall	No. 100	1880	14"	$2,100.00

Type	Pattern	Circa	Dial	Price
Wall	No. 101	1880	12"	$4,000.00

Type	Pattern	Circa	Dial	Price
Wall	No. 69	1880	4½"	$1,250.00

Type	Pattern	Circa	Dial	Price
Wall	No. 123A	1880	4½"	$1,200.00

Type	Pattern	Circa	Dial	Price
Wall	No. 57 Regulator	1880	14"	$3,000.00

Type	Pattern	Circa	Dial	Price
Wall	No. 85 Regulator	1880	14"	$2,500.00

Type	Pattern	Circa	Dial	Price
Wall	No. 70	1880	12" – 16"	$2,200.00

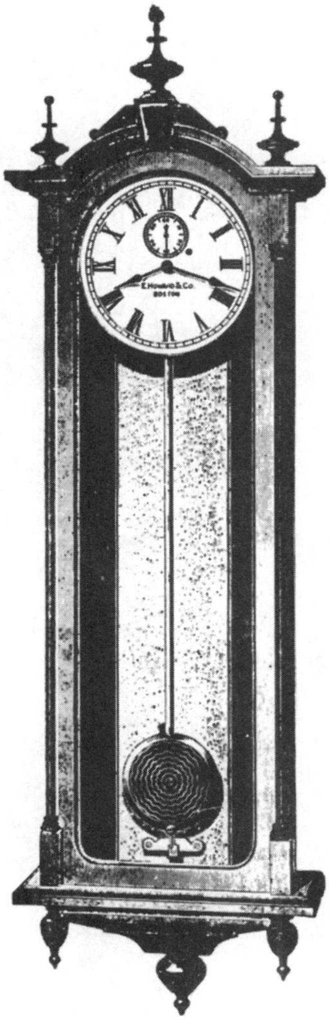

Type	Pattern	Circa	Dial	Price
Wall	No. 71 Regulator	1880	12"	$2,500.00

Howard

Type	Pattern	Circa	Dial	Price
Wall	No. 72 Regulator	1880	12"	$2,500.00

Type	Pattern	Circa	Dial	Price
Wall	No. 86 Regulator (Mercurial Pendulum)	1880	14"	$2,500.00

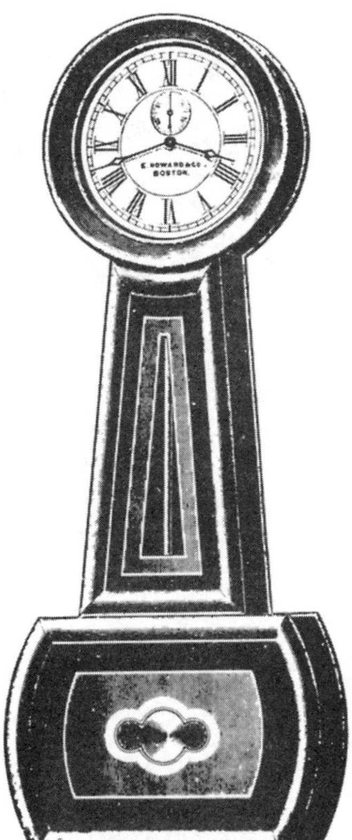

Type	Pattern	Circa	Dial	Price
Wall	No. 1 Regulator	1880	12"	$2,000.00

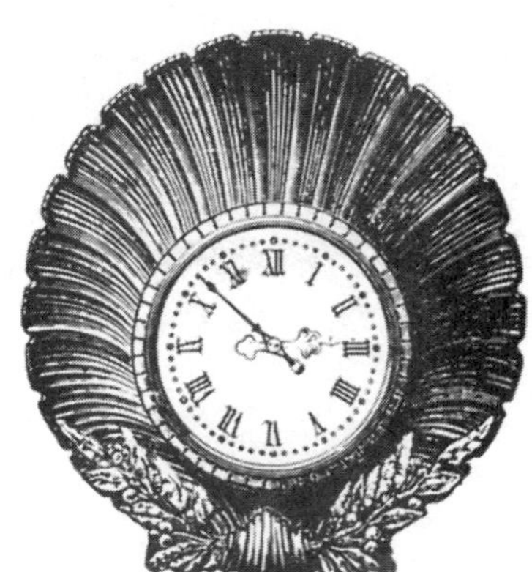

Type	Pattern	Circa	Dial	Price
Wall	No. 99	1880	6"	$300.00

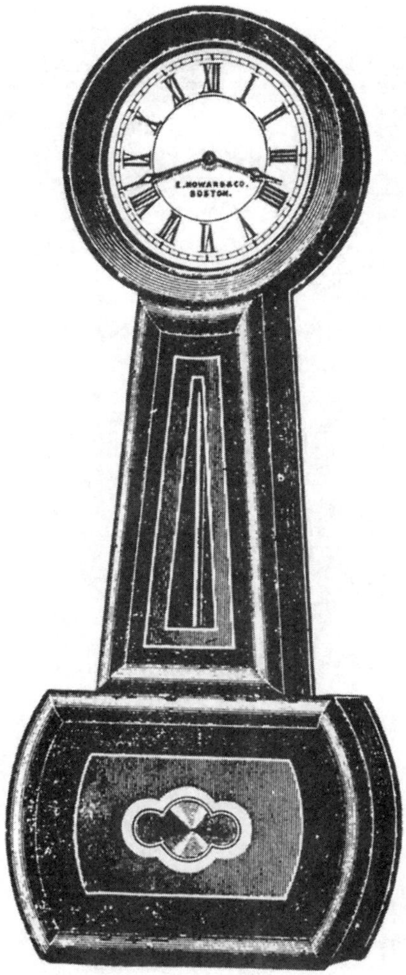

Type	Pattern	Circa	Dial	Price
Wall	No. 2	1880	7" – 10"	$2,200.00

Type	Pattern	Circa	Dial	Price
Wall	No. 89 Regulator	1880	12"	$2,500.00

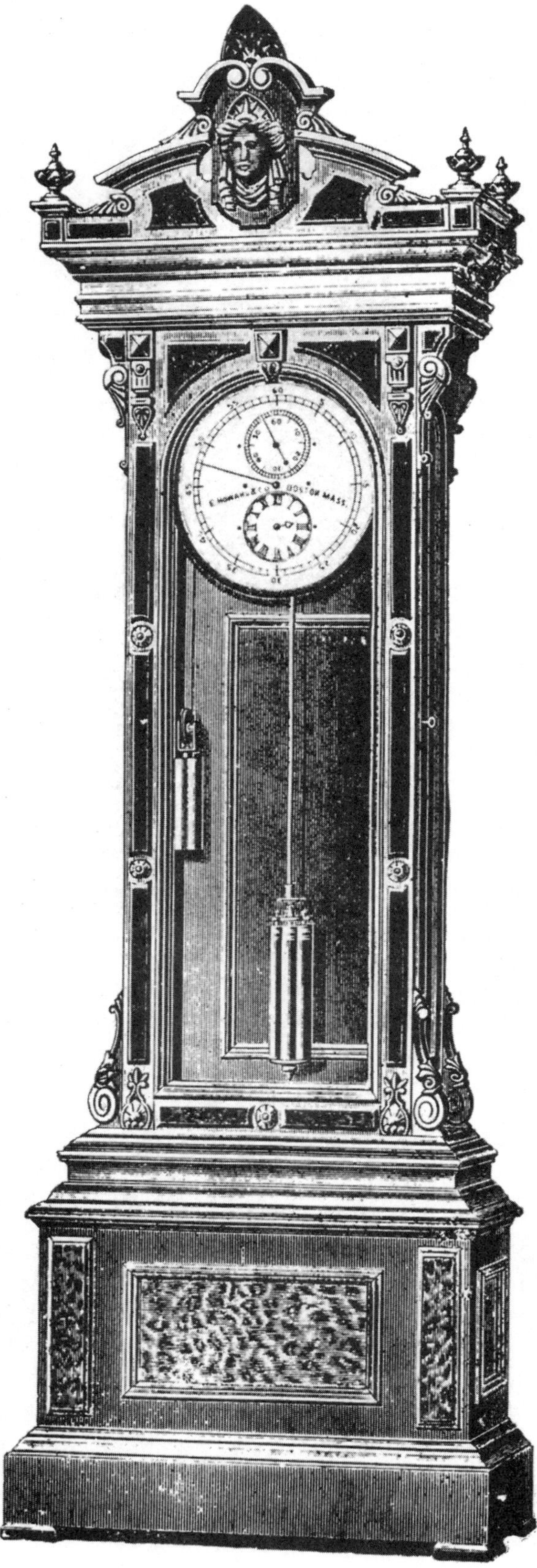

Type	Pattern	Circa	Dial	Price
Hall	No. 68 Regulator	1880	14"	$12,000.00+

Type	Pattern	Circa	Dial	Price
Wall	No. 67 Regulator	1880	15"	$9,000.00+

Type	Pattern	Circa	Dial	Price
Hall	No. 2	1880	16"	$15,000.00+

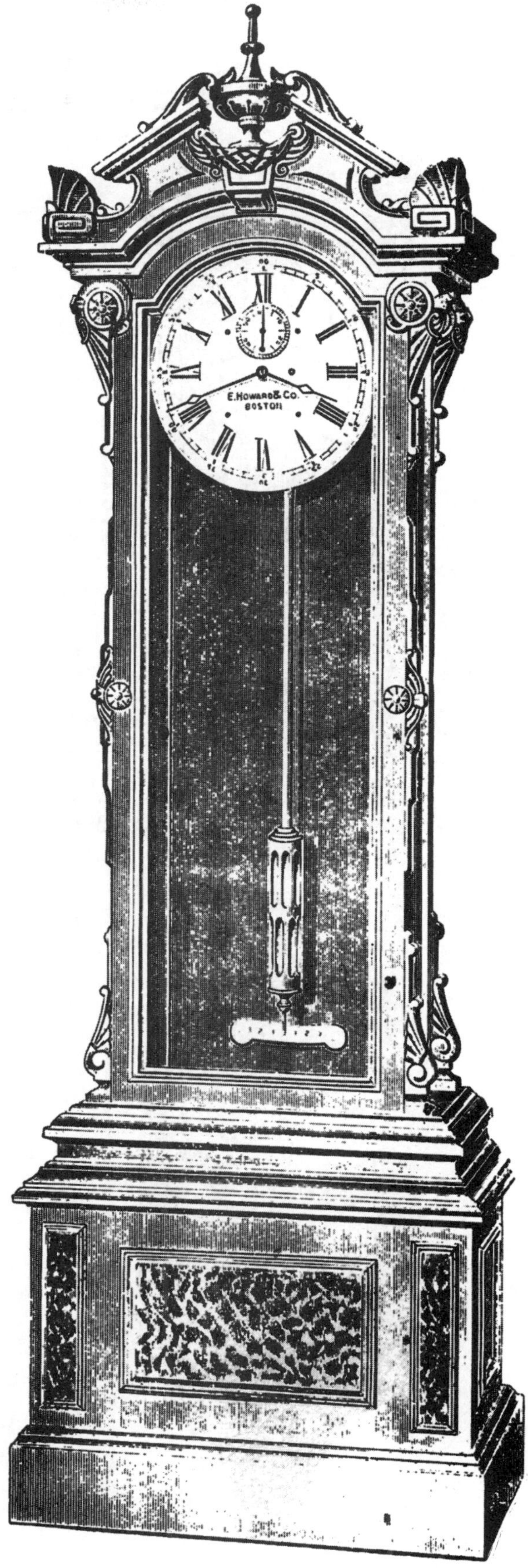

Type	Pattern	Circa	Dial	Price
Hall	No. 61 Regulator	1880	14"	$15,000.00+

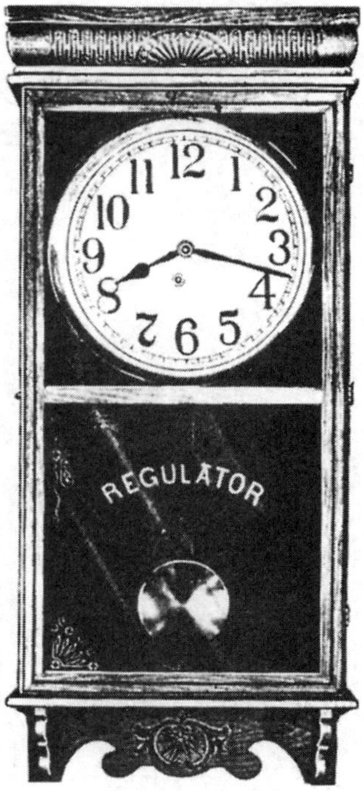

Type	Pattern	Circa	Dial	Price
Wall	Western Union	1930	12"	$425.00

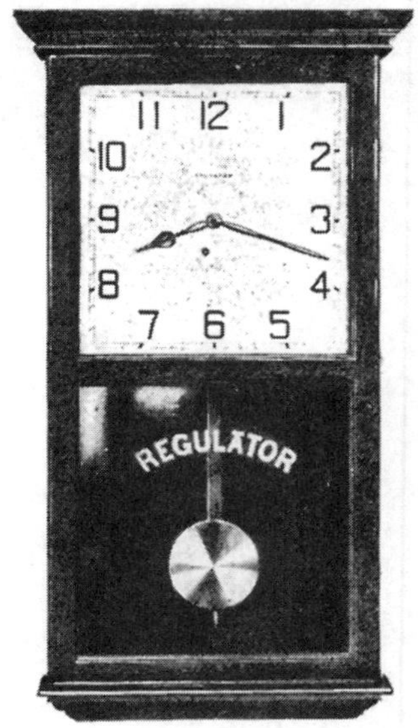

Type	Pattern	Circa	Dial	Price
Wall	Bison	1930	10"	$295.00

Type	Pattern	Circa	Dial	Price
Wall	Nomad	1930	16"	$225.00

Type	Pattern	Circa	Dial	Price
Wall	Natal	1930	16"	$185.00

Type	Pattern	Circa	Height	Price
Banjo	Nordic	1930	26"	$375.00

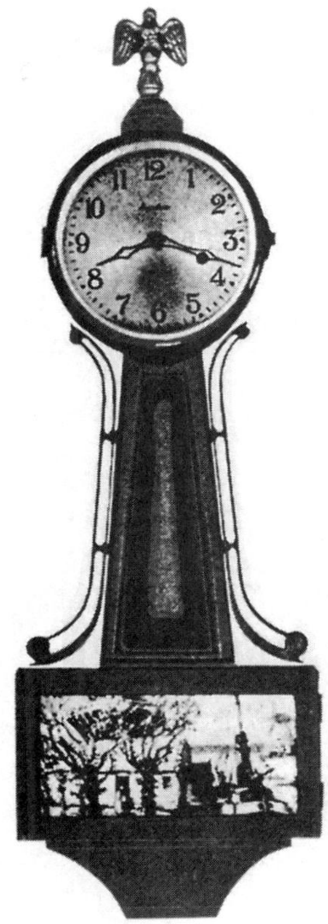

Type	Pattern	Circa	Height	Price
Banjo	Norse	1930	26"	$375.00

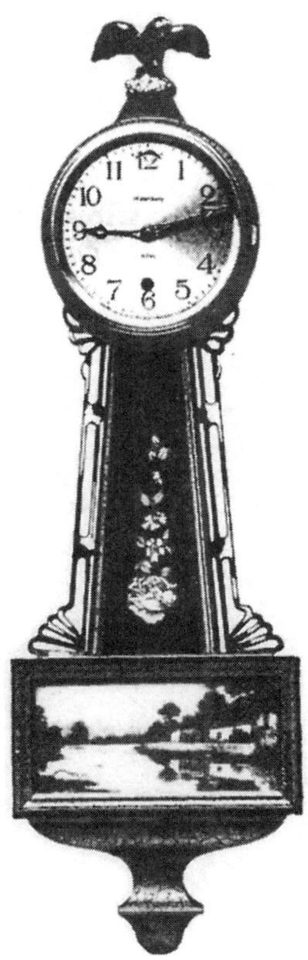

Type	Pattern	Circa	Dial	Price
Banjo	No. 1	1930	6"	$250.00

Type	Pattern	Circa	Dial	Price
Banjo	No. 2	1930	7"	$275.00

Ingraham

Type	Pattern	Circa	Height	Price
Banjo	Treasure	1930	26"	$650.00

Type	Pattern	Circa	Dial	Price
Banjo	Nile	1930	8"	$450.00

Type	Pattern	Circa	Dial	Price
Banjo	Nyanze	1930	8"	$450.00

Type	Pattern	Circa	Dial	Price
Mantel	Hera	1930	7"	$65.00

Type	Pattern	Circa	Dial	Price
Mantel	Magnet	1930	7"	$65.00

Type	Pattern	Circa	Dial	Price
Mantel	Hathor	1930	7"	$65.00

Type	Pattern	Circa	Dial	Price
Mantel	Magic	1930	6"	$65.00

Type	Pattern	Circa	Dial	Price
Mantel	Hermes	1930	6"	$65.00

Type	Pattern	Circa	Dial	Price
Mantel	Meteor	1930	6"	$65.00

Ingraham

Type	Pattern	Circa	Dial	Price
Mantel	Berling	1918	5"	$75.00

Type	Pattern	Circa	Dial	Price
Mantel	Bergundy	1918	5"	$70.00

Type	Pattern	Circa	Dial	Price
Mantel	Grinnell	1918	5"	$80.00

Type	Pattern	Circa	Dial	Price
Mantel	Belmont	1918	5"	$70.00

Type	Pattern	Circa	Dial	Price
Mantel	Hanover	1918	5"	$70.00

Type	Pattern	Circa	Dial	Price
Mantel	Hammond	1918	5"	$80.00

Type	Pattern	Circa	Dial	Price
Mantel	Nomad	1918	5"	$65.00

Type	Pattern	Circa	Dial	Price
Mantel	Argand	1918	5"	$140.00

Type	Pattern	Circa	Dial	Price
Mantel	Arlington	1918	5"	$175.00

Type	Pattern	Circa	Dial	Price
Mantel	Pekin	1918	5"	$140.00

Type	Pattern	Circa	Dial	Price
Mantel	Biltmore	1918	5"	$165.00

Type	Pattern	Circa	Dial	Price
Mantel	Desoto	1918	5"	$195.00

Ingraham

Type	Pattern	Circa	Dial	Price
Mantel	Bonanza	1918	5"	$165.00

Type	Pattern	Circa	Dial	Price
Mantel	Majestic	1918	5"	$175.00

Type	Pattern	Circa	Dial	Price
Mantel	Erna	1900	5"	$200.00

Type	Pattern	Circa	Dial	Price
Mantel	Acme	1900	5"	$160.00

Type	Pattern	Circa	Dial	Price
Mantel	Anchor	1900	5"	$200.00

Type	Pattern	Circa	Dial	Price
Mantel	Beta	1900	5"	$200.00

ESTABLISHED 1829

E. N. WELCH MFG. CO.,

CLOCKS.

Type	Pattern	Circa	Dial	Price
Mantel	Ulmar	1900	5"	$160.00

Type	Pattern	Circa	Dial	Price
Mantel	Zella	1900	5"	$160.00

Type	Pattern	Circa	Dial	Price
Mantel	Viarda, No. 3	1900	5"	$175.00

Type	Pattern	Circa	Dial	Price
Mantel	Shafter	1900	5"	$175.00

Type	Pattern	Circa	Dial	Price
Mantel	Stagno	1900	5"	$175.00

Type	Pattern	Circa	Dial	Price
Mantel	Viarda, No. 2	1900	5"	$180.00

Type	Pattern	Circa	Dial	Price
Mantel	Ursula, No. 3	1900	5"	$180.00

Type	Pattern	Circa	Dial	Price
Mantel	Andree	1900	5"	$175.00

Type	Pattern	Circa	Dial	Price
Mantel	Karina	1900	5"	$175.00

Type	Pattern	Circa	Dial	Price
Mantel	Calve	1900	5"	$180.00

Type	Pattern	Circa	Dial	Price
Mantel	Pinero, No. 1	1900	5"	$180.00

Type	Pattern	Circa	Dial	Price
Mantel	De Merode	1900	5"	$150.00

Type	Pattern	Circa	Dial	Price
Mantel	Pinero, No. 3	1900	5"	$160.00

Type	Pattern	Circa	Dial	Price
Mantel	DeReszke	1900	5"	$160.00

Type	Pattern	Circa	Dial	Price
Mantel	Burkhart	1900	5"	$160.00

Type	Pattern	Circa	Dial	Price
Mantel	Craigen	1900	5"	$150.00

Type	Pattern	Circa	Dial	Price
Mantel	Nansen	1900	5"	$175.00

Type	Pattern	Circa	Dial	Price
Mantel	Alberta	1900	5"	$175.00

Type	Pattern	Circa	Dial	Price
Mantel	Yebba	1900	5"	$150.00

Type	Pattern	Circa	Dial	Price
Mantel	Leno	1900	5"	$145.00

Type	Pattern	Circa	Dial	Price
Mantel	Ursula, No. 1	1900	5"	$180.00

Type	Pattern	Circa	Dial	Price
Mantel	Ursula, No. 2	1900	5"	$180.00

Type	Pattern	Circa	Dial	Price
Mantel	Nicolini A	1900	5"	$180.00

Type	Pattern	Circa	Dial	Price
Mantel	Nicolini B	1900	5"	$180.00

Type	Pattern	Circa	Dial	Price
Mantel	Sorma	1900	5"	$175.00

Type	Pattern	Circa	Dial	Price
Mantel	Viarda, No. 1	1900	5"	$180.00

Type	Pattern	Circa	Dial	Price
Mantel	Roosevelt	1900	5"	$175.00

Type	Pattern	Circa	Dial	Price
Mantel	Belasco	1900	5"	$175.00

Type	Pattern	Circa	Dial	Price
Mantel	Lee	1900	6"	$575.00

Type	Pattern	Circa	Dial	Price
Mantel	Wheeler	1900	6"	$575.00

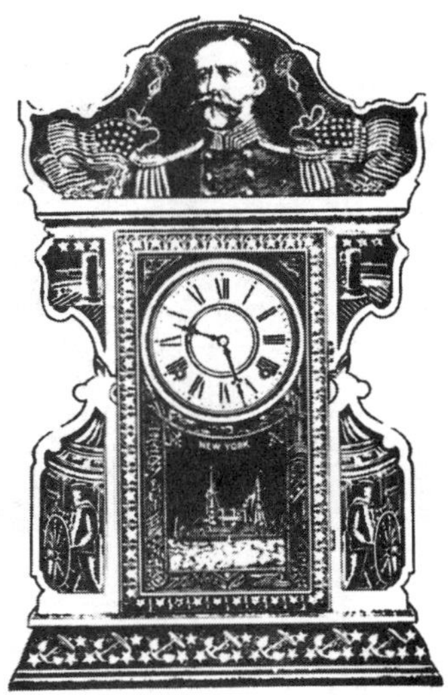

Type	Pattern	Circa	Dial	Price
Mantel	Sampson	1900	6"	$550.00

Type	Pattern	Circa	Dial	Price
Mantel	Schley	1900	6"	$500.00

Type	Pattern	Circa	Dial	Price
Mantel	Dewey	1900	6"	$550.00

Type	Pattern	Circa	Dial	Price
Mantel	The Maine	1900	6"	$550.00

Type	Pattern	Circa	Dial	Price
Cabinet	No. 63	1900	5"	$250.00

Type	Pattern	Circa	Dial	Price
Cabinet	No. 64	1900	5"	$250.00

Type	Pattern	Circa	Dial	Price
Cabinet	No. 65	1900	5"	$250.00

Type	Pattern	Circa	Dial	Price
Cabinet	No. 66	1900	5"	$250.00

Type	Pattern	Circa	Dial	Price
Cabinet	No. 61	1900	5"	$250.00

Type	Pattern	Circa	Dial	Price
Cabinet	No. 62	1900	5"	$250.00

Type	Pattern	Circa	Dial	Price
Wall	Star Pointer Calendar	1900	12"	$950.00

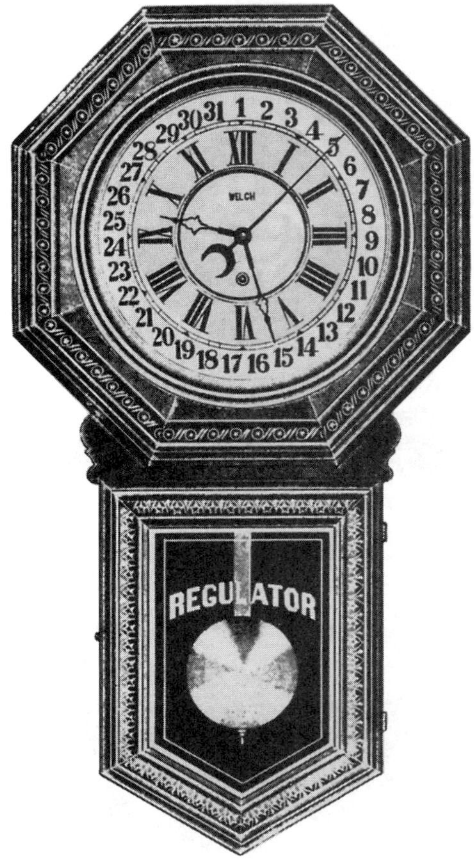

Type	Pattern	Circa	Dial	Price
Wall	St. Clair Calendar	1900	12"	$800.00

Type	Pattern	Circa	Dial	Price
Wall	Patchen	1900	12"	$600.00

Type	Pattern	Circa	Dial	Price
Wall	Patchen Calendar	1900	12"	$650.00

Type	Pattern	Circa	Dial	Price
Wall	No. 11 Regulator	1900	10"	$3,400.00

Type	Pattern	Circa	Dial	Price
Wall	No. 8 Regulator	1900	9"	$3,400.00

Type	Pattern	Circa	Dial	Price
Wall	Gentry	1900	12"	$800.00

Type	Pattern	Circa	Dial	Price
Wall	Gentry Calendar	1900	12"	$900.00

Type	Pattern	Circa	Dial	Price
Wall	Star Pointer	1900	12"	$900.00

Type	Pattern	Circa	Dial	Price
Wall	Gallery	1900	18" and 24"	$1,200.00

Type	Pattern	Circa	Dial	Price
Wall	No. 12 Regulator	1900	12"	$3,750.00

ITHACA CLOCK CO.

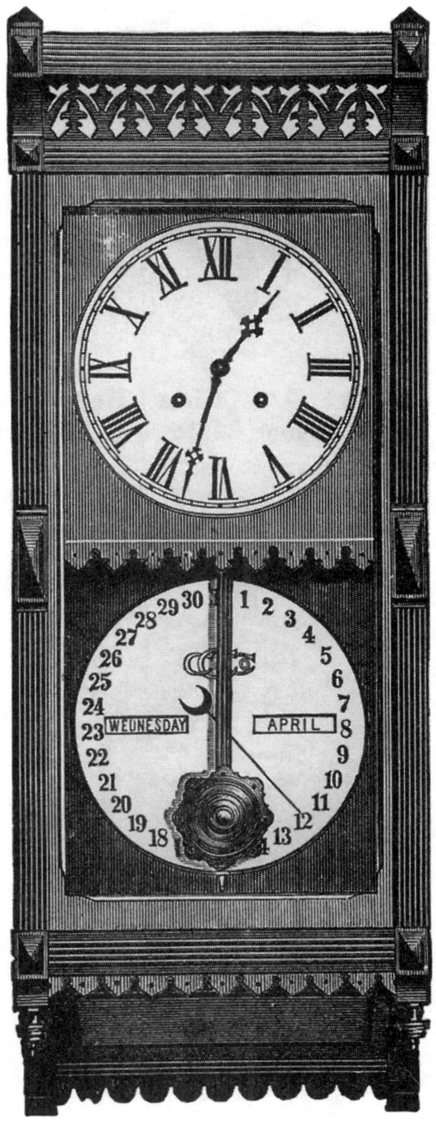

Type	Pattern	Circa	Dial	Price
Calendar	Brisbane	1880	11"	$3,200.00

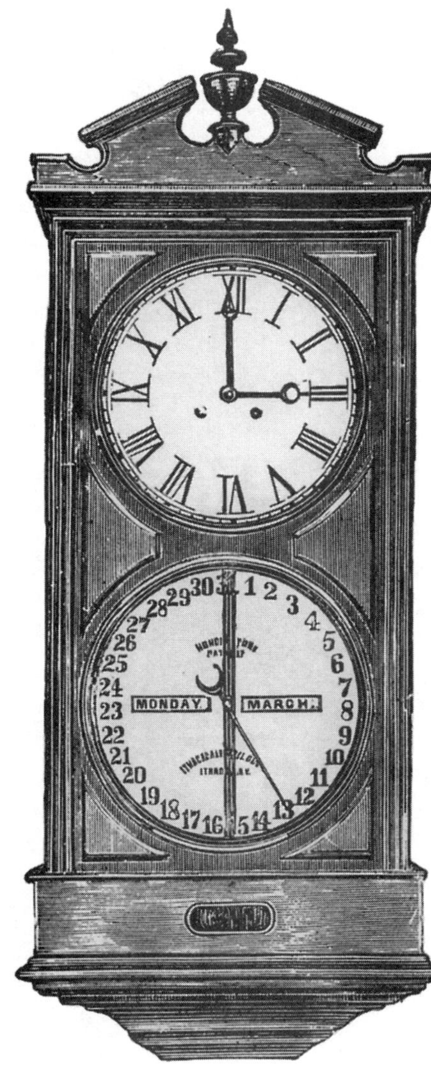

Type	Pattern	Circa	Dial	Price
Calendar	No. 2 Regulator	1880	12"	$3,000.00

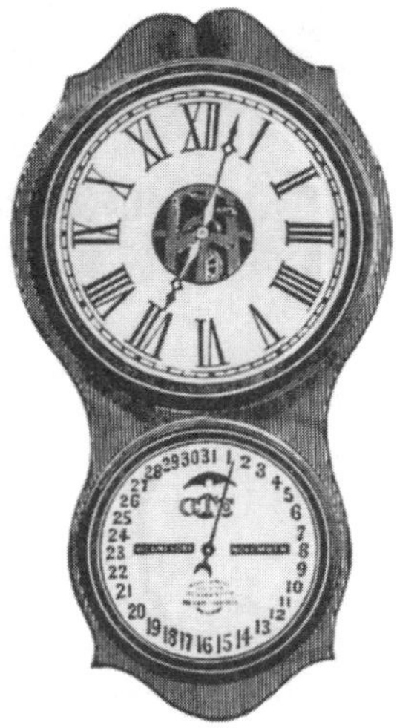

Type	Pattern	Circa	Height	Price
Calendar	No. 4 Hanging Office	1880	28"	$1,500.00

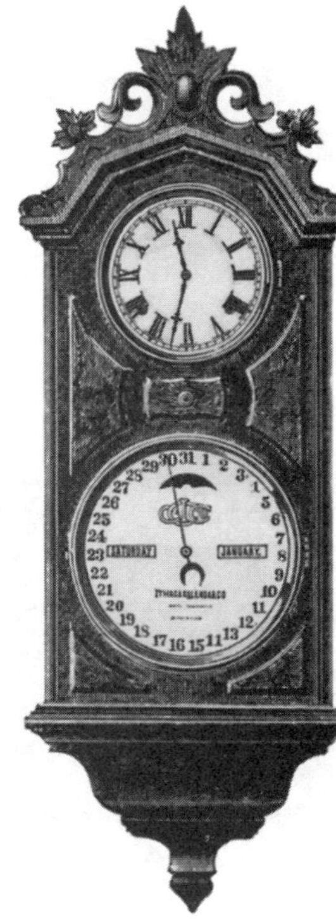

Type	Pattern	Circa	Height	Price
Calendar	No. 6 Hanging Library	1880	32"	$2,400.00

Type	Pattern	Circa	Height	Price
Calendar	No. 5 Emerald	1880	33"	$1,500.00

Type	Pattern	Circa	Height	Price
Calendar	No. 7 Hanging Cottage	1880	29"	$2,400.00

Type	Pattern	Circa	Height	Price
Calendar	No. 8 Shelf Library	1880	26"	$850.00+

Type	Pattern	Circa	Height	Price
Calendar	No. 10 Farmer's	1880	25"	$875.00

Type	Pattern	Circa	Height	Price
Calendar	No. 9 Shelf Cottage	1880	23"	$1,500.00

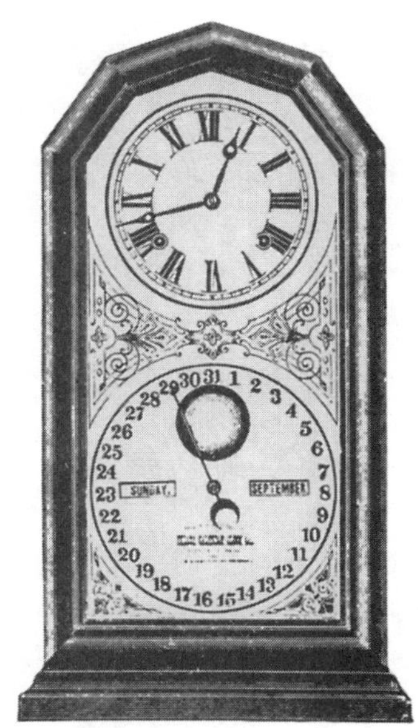

Type	Pattern	Circa	Height	Price
Calendar	No. 11 Octagon	1880	21"	$825.00

Ithaca

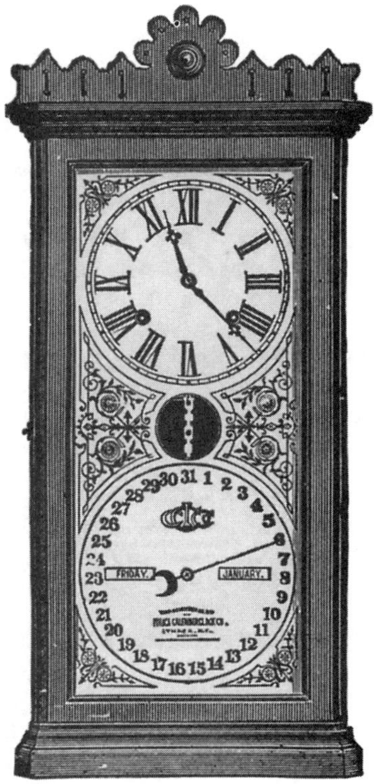

Type	Pattern	Circa	Dial	Price
Calendar	Granger	1880	8"	$2,200.00

Type	Pattern	Circa	Height	Price
Calendar	No. 0 Bank	1880	61"	$5,500.00+

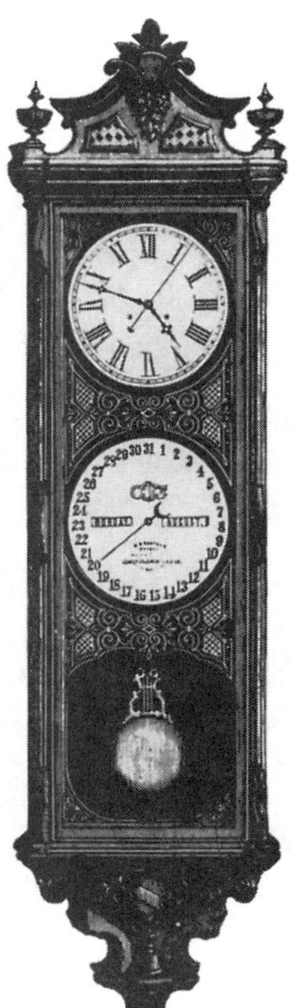

Type	Pattern	Circa	Height	Price
Calendar	No. 1 Regulator	1880	72"	$5,000.00

Type	Pattern	Circa	Height	Price
Calendar	No. 2 Bank	1880	61"	$5,200.00

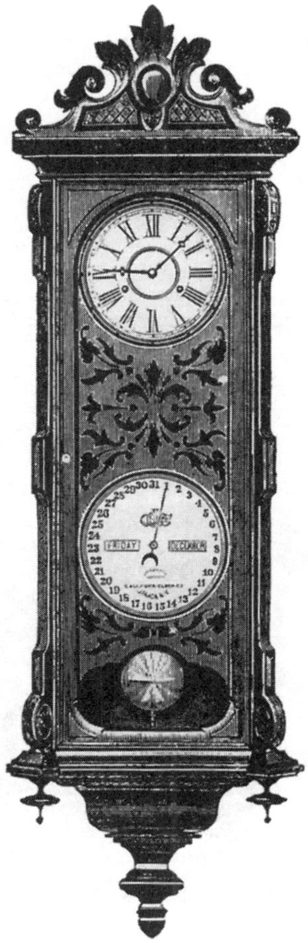

Type	Pattern	Circa	Height	Price
Calendar	No. 3 Vienna	1880	52"	$6,500.00+

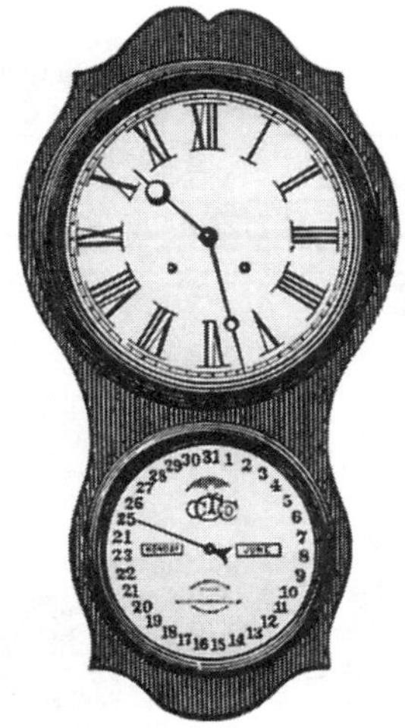

Type	Pattern	Circa	Height	Price
Calendar	No. 4 Hanging Office	1880	28"	$1,450.00

Type	Pattern	Circa	Height	Price
Calendar	No. 3½ Parlor	1880	20"	$3,750.00

Type	Pattern	Circa	Height	Price
Calendar	No. 4½ Favorite	1880	32"	$2,750.00

Sempire Clock Co.

SELF-WINDING

Type	Pattern	Circa	Dial	Price
Wall	No. 65 Jewelers	1900	12"	$3,500.00

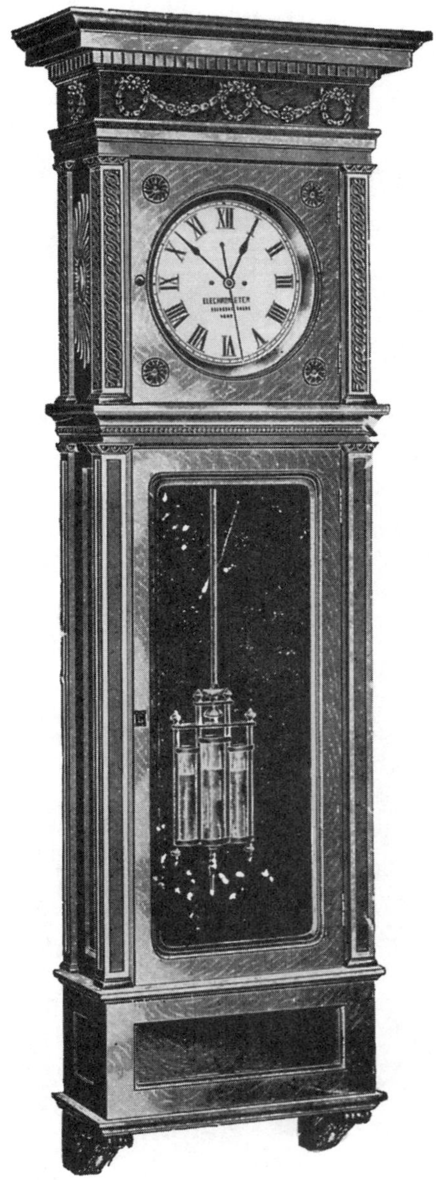

Type	Pattern	Circa	Dial	Price
Wall	Palesburg	1900	12"	$3,500.00

Type	Pattern	Circa	Dial	Price
Wall	No. 22	1900	12"	$2,400.00

Type	Pattern	Circa	Dial	Price
Wall	No. 58 Jewelers	1900	12"	$2,400.00

Type	Pattern	Circa	Dial	Price
Wall	No. 48 Vestibule	1900	12"	$3,000.00

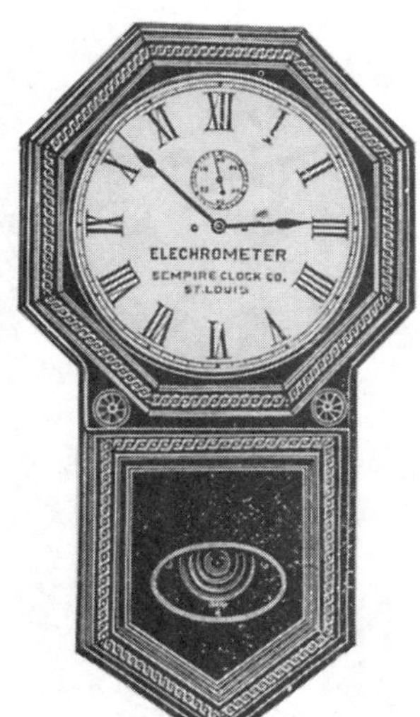

Type	Pattern	Circa	Dial	Price
Wall	No. 9 School	1900	12"	$1,500.00+

Type	Pattern	Circa	Dial	Price
Wall	No. 32 Office	1900	12"	$950.00

Type	Pattern	Circa	Dial	Price
Wall	No. 68 Jewelers	1900	12"	$3,500.00

Type	Pattern	Circa	Dial	Price
Wall	No. 36 Gallery	1900	18"	$975.00

Type	Pattern	Circa	Dial	Price
Wall	No. 8 Gallery	1900	14"	$850.00+

WM. L. GILBERT CLOCK CO.

ESTABLISHED 1825

Type	Pattern	Circa	Dial	Price
Mantel	Bonanza	1900	5½"	$200.00

Type	Pattern	Circa	Dial	Price
Mantel	Major	1900	5½"	$200.00

Type	Pattern	Circa	Dial	Price
Mantel	Arena	1900	5½"	$225.00

Type	Pattern	Circa	Dial	Price
Mantel	Melody	1900	5½"	$200.00

Type	Pattern	Circa	Dial	Price
Mantel	Astor	1900	5½"	$215.00

Type	Pattern	Circa	Dial	Price
Mantel	Palisade	1900	5½"	$195.00

Type	Pattern	Circa	Dial	Price
Mantel	Shannon	1900	5½"	$180.00

Type	Pattern	Circa	Dial	Price
Mantel	Victor	1900	5½"	$175.00

Type	Pattern	Circa	Dial	Price
Mantel	Pinta	1900	5½"	$180.00

Type	Pattern	Circa	Dial	Price
Mantel	Norwood	1900	5½"	$180.00

Type	Pattern	Circa	Dial	Price
Mantel	Eton	1900	5½"	$175.00

Type	Pattern	Circa	Dial	Price
Mantel	Westin	1900	5½"	$180.00

Type	Pattern	Circa	Dial	Price
Mantel	Sumner	1900	5½"	$175.00

Type	Pattern	Circa	Dial	Price
Mantel	Pauline	1900	5½"	$175.00

Type	Pattern	Circa	Dial	Price
Mantel	Cortez	1900	5½"	$175.00

Type	Pattern	Circa	Dial	Price
Mantel	Helma	1900	5½"	$175.00

Type	Pattern	Circa	Dial	Price
Mantel	Congo	1900	5½"	$180.00

Type	Pattern	Circa	Dial	Price
Mantel	Jamaica	1900	5½"	$180.00

Type	Pattern	Circa	Dial	Price
Mantel	Laredo	1900	5½"	$185.00

Type	Pattern	Circa	Dial	Price
Mantel	Tuxedo	1900	5½"	$180.00

Type	Pattern	Circa	Dial	Price
Mantel	Nina	1900	5½"	$185.00

Type	Pattern	Circa	Dial	Price
Mantel	Windsor	1900	5½"	$175.00

Type	Pattern	Circa	Dial	Price
Mantel	Pocono	1900	5½"	$185.00

Type	Pattern	Circa	Dial	Price
Mantel	Saranac	1900	5½"	$185.00

Type	Pattern	Circa	Dial	Price
Mantel	Grenada	1900	5½"	$180.00

Type	Pattern	Circa	Dial	Price
Mantel	Arlington	1900	5½"	$190.00

Type	Pattern	Circa	Dial	Price
Mantel	National	1900	5½"	$185.00

Type	Pattern	Circa	Dial	Price
Mantel	Colonia	1900	5½"	$185.00

Type	Pattern	Circa	Dial	Price
Mantel	Savoy	1900	5½"	$185.00

Type	Pattern	Circa	Dial	Price
Mantel	Breslin	1900	5½"	$185.00

Type	Pattern	Circa	Dial	Price
Mantel	Laurell	1900	6"	$420.00

Type	Pattern	Circa	Dial	Price
Mantel	Geranium	1900	6"	$400.00

Type	Pattern	Circa	Dial	Price
Mantel	Ersa	1880	6"	$395.00

Type	Pattern	Circa	Dial	Price
Mantel	Mahuta	1880	6"	$375.00

Type	Pattern	Circa	Dial	Price
Mantel	Prince	1800	6"	$395.00

Type	Pattern	Circa	Dial	Price
Mantel	Dio	1880	6"	$395.00

Type	Pattern	Circa	Dial	Price
Mantel	Eros	1880	6"	$395.00

Type	Pattern	Circa	Dial	Price
Mantel	Calpe	1880	6"	$385.00

Type	Pattern	Circa	Dial	Price
Mantel	Advance	1900	6"	$250.00

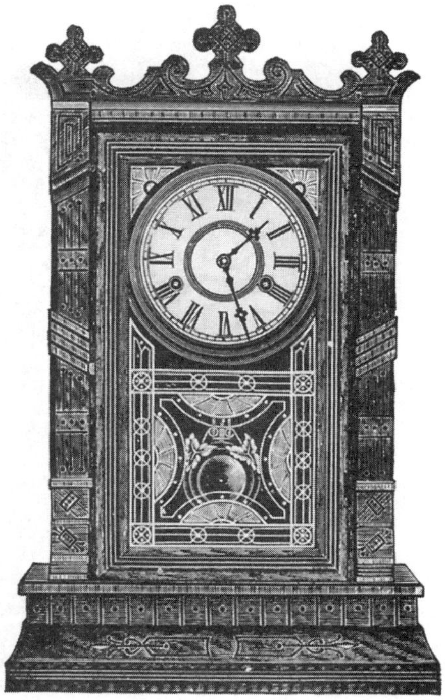

Type	Pattern	Circa	Dial	Price
Mantel	Hestia	1880	6"	$350.00

Type	Pattern	Circa	Dial	Price
Mantel	Bishop	1900	6"	$475.00

Type	Pattern	Circa	Dial	Price
Mantel	Queen	1900	6"	$450.00

Type	Pattern	Circa	Dial	Price
Mantel	Concord	1900	6"	$400.00

Type	Pattern	Circa	Dial	Price
Mantel	Eclipse	1900	6"	$475.00

Type	Pattern	Circa	Dial	Price
Mantel	Perfect	1900	6"	$525.00

Type	Pattern	Circa	Dial	Price
Mantel	Occidental	1900	6"	$550.00

Type	Pattern	Circa	Dial	Price
Wall	Admiral	1900	12"	$950.00

Type	Pattern	Circa	Dial	Price
Wall	Regulator B	1900	12"	$975.00

Type	Pattern	Circa	Dial	Price
Wall	Consort	1900	12"	$500.00

Type	Pattern	Circa	Dial	Price
Wall	Office Drop Calendar	1900	15"	$1,750.00

Gilbert

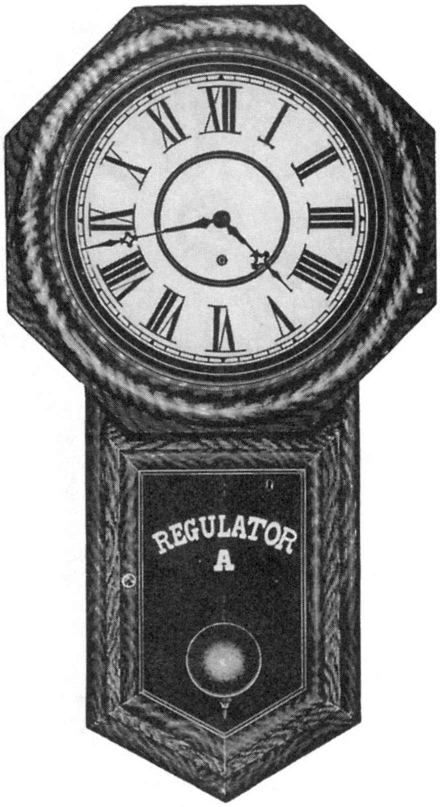

Type	Pattern	Circa	Dial	Price
Wall	Regulator A	1900	12"	$475.00

Type	Pattern	Circa	Dial	Price
Wall	Regulator 10	1880	10"	$2,500.00

Type	Pattern	Circa	Dial	Price
Wall	Drexel	1900	6"	$495.00

Type	Pattern	Circa	Dial	Price
Wall	Alpine	1900	12"	$400.00

Type	Pattern	Circa	Dial	Price
Wall	Circle Drop	1900	12"	$475.00

Type	Pattern	Circa	Dial	Price
Wall	Thespian	1900	8"	$1,750.00

Type	Pattern	Circa	Dial	Price
Wall	Columbia	1900	8"	$1,900.00

Schroeder's ANTIQUES Price Guide

. . . is the #1 bestselling antiques & collectibles value guide on the market today, and here's why . . .

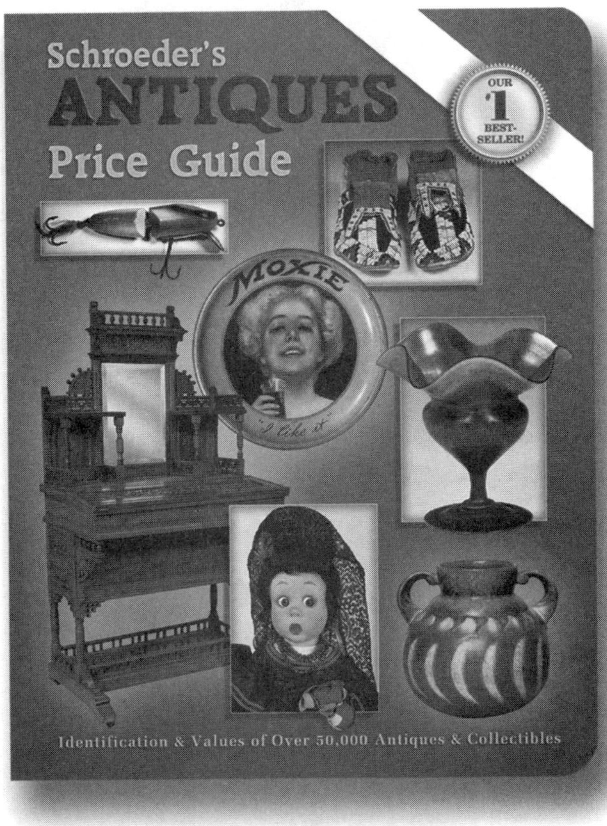

8½ x 11, 608 Pages, $14.95

- *More than 400 advisors, well-known dealers, and top-notch collectors work together with our editors to bring you accurate information regarding pricing and identification.*

- *More than 50,000 items in over 600 categories are listed along with hundreds of sharp original photos that illustrate not only the rare and unusual, but the common, popular collectibles as well.*

- *Each large close-up shot shows important details clearly. Every subject is represented with histories and background information, a feature not found in any of our competitors' publications.*

- *Our editors keep abreast of newly developing trends, often adding several new categories a year as the need arises.*

If it merits the interest of today's collector, you'll find it in *Schroeder's*. And you can feel confident that the information we publish is up to date and accurate. Our advisors thoroughly check each category to spot inconsistencies, listings that may not be entirely reflective of market dealings, and lines too vague to be of merit. Only the best of the lot remains for publication.

Collector Books
P.O. Box 3009
Paducah, KY 42002-3009
1-800-626-5420
www.collectorbooks.com

COLLECTOR BOOKS

A Division of Schroeder Publishing Co., Inc.